TANK CAR

Color Guide

Volume 2: Stub Sill Cars

by **James Kinkaid**

Library of Congress
Catalog Card No. 2010928147

First Printing
ISBN 1-58248-322-1

MORNING SUN BOOKS INC.

Robert J. Yanosey, President
To access our full library In Color visit us at
www.morningsunbooks.com

Published by
Morning Sun Books, Inc.
9 Pheasant Lane
Scotch Plains, NJ 07076
Printed in Korea

Dedication

I would like to dedicate this book to Eric Neubauer. Eric has been accumulating car production information for many years and if anybody can be described as a past master on freight car roster, lot and production information, he can. He has been always willing to help me out on every single Morning Sun color guide that I have authored and it is no exaggeration to state that his assistance has made my work results infinitely better than they would have been otherwise. So, even though I have said it many times to him, it needs to be said again: *Thanks,* Eric!

Acknowledgements

As with all Morning Sun Color Guides, this volume would not be possible without the help of many contributors. This is even more so since these two books rely heavily on individual collections. So, I offer my heartfelt thanks to: Gib Allbach, Calvin T. Banse, G. Bolinsky, Craig Bossler, David Casdorph, David Graeff, Emery Gulash, Matt Herson, David Hickcox, Pat Holden, Al Lanier, Dave McKay, Dave Nelson, Eric Neubauer, Jim Rogers, William Rosenberg, Lou Schmitz, Wade Stevenson, Jim Thorington, Ed Trundle, George Turnbull, Bernie Wooller, Paul C. Winters, Michael Wise, Lee Yoder and Chuck Yungkurth of Rail Data Service. And again, my many thanks to Bob Yanosey of Morning Sun Books for his help, advice and patience with me on these two volumes.

Table of Contents

TANK CAR Color Guide

Volume 2: Stub Sill Cars

Introduction

This is Volume Two of a two volume set and as such, they are meant to be utilized together. It presents tank cars constructed without full center sills, also known as stub sill cars. Volume One deals with tank cars equipped with full underframe center sills.

Tank cars built without full center sills are certainly nothing new: Van Dyke had built a number of cars to this design in the late 1880's. However, the concept never went anywhere and it was not until the advent of the high capacity tank car around 1962 that the design feature was reintroduced. Most modern tank cars are built this way, with the exceptions of those constructed with aluminum tanks or the rather rare cars that require full length center sills due to specification requirements.

Tank Car Specifications

The Master Car Builders' Association first issued rules designed to formalize tank car repairs and manufacturing in 1903. As early as January 1918 they even issued a specification for all-welded cars for chlorine and propane use.

On February 1, 1920 the American Railway Association (ARA) took over this effort, and carried onwards with regulations until the advent of the Interstate Commerce Commission (ICC). Typically MCB cars were simply restenciled with ARA markings although the ARA did develop several entirely new sets of specifications.

The ICC took over jurisdiction on July 1, 1927 for those cars utilized in hazardous commodity service while the Association of American Railroads (AAR) issued separate specifications for cars used strictly in non-hazardous service. At this point in time the number of specifications began to greatly increase, reflecting the greater desire to better suit the tank car to the commodity.

During World War II, three sets of temporary specifications (USG-A, -B and -C) were issued to allow construction of tank cars for crude oil and gasoline transport using thinner steels.

On April 1, 1967 the United States Department of Transportation superceded the ICC in regulating hazardous commodity transport and all those specifications issued after that date came marked "DOT." In general, cars that previously had ICC markings were simply restenciled with the DOT prefix. The AAR continued to offer specifications for strictly non-hazardous use cars.

All tank cars are marked with a specification stenciled on each side of the tank. A modern specification consists of an alphabetic prefix followed by a series of numerals and/or letters. The alphabetic prefix denotes the agency under whose specifications the tank was constructed (i.e., ICC, DOT or AAR). The following three numerals designate the class of the car and any following numerals or letters designate such things as the tank test pressure, construction material, special protective systems (such as thermal jackets) and type of construction. While a specification rarely indicates the shape of the car it can indicate a number of things. Let's take a typical modern specification: DOT111A100W1. This means that it was issued by the DOT, it is a class 111 design (which signifies a general service car lacking a dome), the "A" in this instance is only a place holder (but on some specifications implies a meaning), the 100 indicates the tank working pressure, the "W" signifies fusion welding and the suffix "1" denotes a car that might (or might not) be insulated and has a bottom outlet (plus other features).

Tank Car Designs

Most tank car tanks are constructed of steel but many aluminum tanks also exist as well as a limited number of nickel and stainless steel tanks. At one time there were also wooden tanks. Depending on the specification cars may be non-insulated or have some kind of insulation (typically enclosed within an outer thin walled jacket) or be otherwise thermally protected. Cars may be top loaded only, top or bottom loaded or bottom loaded only. They may have heater pipes or coils, which typically use steam to liquefy ladings that might have solidified during transit or to make thick commodities easier and quicker to unload. Cars may be unlined or have a variety of interior linings to protect the tank from the lading, or vice versa. They may have sparger systems installed (which introduce a fluid into the commodity to help unload it) or in the case of the covered hopper designs, have fluidized or pneumatic discharge features.

Tank cars may be grouped into three basic categories: (A) non-pressure (or general service) designs, (B) pressure cars and (C) special service or limited application designs.

Non-pressure cars are those with tank working pressures of 100 psi or less and may have multiple compartments, each of which is its own separate reservoir with attendant manways (or domes), safety valves or vents, and outlets. Considered general service cars, they may be used to transport either liquid or solid (loaded and unloaded as liquids) commodities as well as some liquefied compressed gasses. The ladings may be either non-hazardous (or non-regulated) or hazardous (meaning that their transport is regulated by government authority). General service cars may have expansion domes. Typical specifications for this design would be MCB III, ARA III, the USG cars, ICC/DOT103, ICC/DOT104, AAR201 and AAR203, all of which have expansion domes which allowed for fluid expansion and entry way purposes.

By the 1960's tank cars began to be built without domes since the new designs allowed for product expansion within the tank itself. These cars thus utilized simple "manways." Typical specifications for non-pressure cars would be ICC109, the entire range of ICC/DOT111 cars and AAR205.

Pressure cars are almost totally used in hazardous commodity transportation, which can be either in liquid, compressed gas or solid form. These cars have tank working pressures above 100 psi and the tank working pressure will be part of the specification. In addition ICC/DOT cars will have water capacity markings. Typical specifications for these cars might be MCB V, ARA V, ICC/DOT105, ICC/DOT109, ICC/DOT112 and related 114 specifications. None of these cars have domes but rather manways, often covered with hoods (called bonnets).

Special service cars are just about all of the exotic cars seen. These include the helium cars (ICC/DOT107), the "true" AAR class TMU cars (which have field removable tanks available for road carriage) such as ICC/DOT106 and ICC/DOT110 and the wooden tank cars (ICC108). There are also the cryogenic cars that utilize vacuum as part of the insulation process which include specifications such as ICC/DOT113 and AAR204. Then there are the super-insulated cars that utilize regular insulation combined with special construction features built under specification AAR206. And finally there are the dry commodity cars that typically resemble covered hoppers. Those are constructed under AAR specification AAR207.

Single Compartment Non-Insulated Cars

ACFX 17382
DOT114A340W
▲ Note the yellow manway and the mud guards over the wheels on this car found at Calvert City, Kentucky on March 22, 1984. ACF constructed it in October 1968. It was part of the four car set 17381-17384.
(Al Lanier, Jim Kinkaid collection)

ACFX 72727
DOT111A100W1
► A car in Joliet, Illinois on June 16, 1991. At the time it was under lease to Union Carbide and was a 30,096 gallon car assembled by American Car and Foundry in February 1989 under lot 71-14355. As such it was one of 32 cars assigned the serial numbers 72699-72730, built from August 1988 through February 1989. *(Gib Albach)*

ACFX 77601
▼ A car on the move at Baltimore, Maryland in September 1991. It was constructed in February 1976 and part of a four car set numbered as ACFX 77600-77603. These cars were furnished under builder's lot 71-15207. *(Jim Rogers)*

ACFX 83108
DOT111A100W1
▲ At Twin Falls, Idaho on July 20, 1972. This car was manufactured in October 1969 by ACF's Milton, Pennsylvania plant. It was part of the 15-car set numbered 83094-83108. "Inedible" refers to the type of rendered by-products from packing plants. *(Jim Kinkaid collection)*

ACFX 88896
DOT111A100W
◀ A car underway in April 1971. Assembled in December 1968 it was part of a 15-car set numbered 88887-88901. Of interest is the builder's logo painted on the bolster. *(William Rosenberg)*

ACFX 88930
ICC111A100W5
▼ Found and placed on film in October 1969, it had just been built the prior January. In the equipment register it was listed as a single car and it had a capacity of 20,452 gallons. *(Paul C. Winters)*

AMIX 300082
DOT111A100W1
▼ At Kansas City, Missouri when photographed in 1996 it has an October 1994 manufacture date, constructed by Trinity Industries' Tulsa, Oklahoma plant. The car was part of the 27-car series numbered 300075-300101 which were assembled under file 9410. The reporting mark for this 30,168 gallon car was assigned to Railcar Leasing Specialists, Incorporated while the logo is that of Minnesota Corn Processors, of Marshall, Minnesota. *(Jim Kinkaid)*

CELX 8334
DOT111A100W1
▲ Our first Celtran car in Volume Two is shown in this illustration taken at Wellington, Kansas in December 1989. The reinforcement rings are of note on this car, which was built in August 1972. It belonged to the series 8300-8334 which had been assembled at Union Tank Car in East Chicago in mid-1972. *(Jim Kinkaid)*

CELX 11261
DOT111A100W1
▼ A car seen in June 1990 at Altoona, Pennsylvania. It was a part of the Celtran series 11250-11288. Those cars, constructed under Richmond Tank Car's lot 2206 at Houston, Texas, were delivered in April through June 1974. *(Jim Kinkaid)*

CGTX 20276
DOT111A100W1
▲ General American Transportation's Sharon, Pennsylvania plant built this car in January 1975 for owner Canadian General Transit Company, Limited and it was put on film at South Edmonton, Alberta in Canada on June 23, 1982. In the register it was shown as part of the set 20274-20280 although more cars may have been involved. The lessee was a company originally founded in 1947 as Leduc Calmar Oil Company, of Alberta. *(Matt Herson, Jim Kinkaid collection)*

CHRX 1001
ICC112A340W
▼ The photographer caught this car in Worcester, Massachusetts on March 16, 1971. It had been built by the Canadian builder Hawker Siddley at its Trenton, Nova Scotia plant in June 1968. CHRX was the reporting mark assigned to Canadian Hidrogas Resources, Limited and the 34,000 gallon car was part of the 1001-1008 series. *(George Turnbull, Jim Kinkaid collection)*

CMGX 13043
DOT111A100W2
▲ Cities Service Company (a company that can be traced back to 1910) owned the car shown here. It was part of the 210-car group 13000-13209, which were 13,300 gallon cars. This one was constructed in January 1972 and photographed that March.
(Rail Data Service, Jim Kinkaid collection)

DACX 1559
DOT103W
▲ Placed on film in July 1985 the car was owned by the Darling and Company. An 8,000 gallon capacity car fitted with heater coils, it was part of the DACX series 1551-1560. Constructed by General American Transportation Company in early 1965 it was one of the relatively few cars built with stub sills fitted with true expansion domes. *(Rail Data Service, Jim Kinkaid collection)*

DODX 14024
DOT111A100W1
▲ An interesting car! Assembled in August 1966 as part of the 200-car set 14000-14199, it was then converted in April 1976 from a W2 to a W1 specification. The change (four cars were so done) allowed the car to be fitted with a bottom outlet. It was a 125-ton capacity car that had end-of-car cushioning: very rare for tank cars. The photo of this 20,000 gallon car was taken in June 1984.
(Jim Kinkaid collection)

DODX 14332
DOT111A100W1
◀ Brand new, its photo was taken on February 28, 1976 at Muhlenburg Township, Pennsylvania while a string of these cars was being delivered. Assembled by American Car and Foundry that month, it was one of 162 numbered DODX 14200-14361. These were also 20,000 gallon cars. *(Craig Bossler)*

DOWX 3318
▼ DOT112A340W
Used in vinyl chloride service it was at Reading, Pennsylvania on June 22, 1969, having been constructed in April 1967 by General American Transportation. The car, which was of 31,388 gallon capacity, was part of a five car group numbered as 3314-3318. *(Craig Bossler)*

DUPX 11402
DOT112S200W
▲ This DuPont example, found in Reading, Pennsylvania on February 10, 1988, illustrates a car fitted with head shields, as delineated by the "S" within the specification. A close look will show where the head shield has been welded to the end of the tank shell. It was fabricated in January 1987 by Union Tank Car at East Chicago and one of six built as 11401-11406. *(Craig Bossler)*

DUPX 17370
DOT111A100W5
▼ We're seeing a car drifting downgrade at Altoona, Pennsylvania in June 1990. Since the car is carrying ferric chloride, it was probably painted brown to reduce the staining (hints of which can be seen). The 20,800 gallon car was part of a 36-car set numbered as 17361-17396. Rubber lined, they were assembled by General American in December 1975 and January 1976. *(Jim Kinkaid)*

DUPX 20152
ICC114A340W
▶ Supplied by General American Transportation Company in May 1968 the car was photographed in July 1969. It was a 20,200 gallon car assigned into the 20150-20179 set. Cars were also furnished in June of that year.
(William Rosenberg)

DUPX 28001
ICC114A400W
▼ March 22, 1984 was when this photograph was taken at Paducah, Kentucky. Apparently constructed in January 1966, the 28,300 gallon car was part of the number series 28001-28029.
(Al Lanier, Jim Kinkaid collection)

DUPX 29153
ICC111A100W1
▼ Probably assembled in late 1968 the car was part of the 30-car set numbered 29131-29160. The 30,032 gallon car was at Sheffield, Alabama on November 17, 1983 while in methanol service. *(Al Lanier, Jim Kinkaid collection)*

GATX 1801
ICC111A100W5
▲ Here is a car found in May 1968. Built in May 1965 the car was rubber lined. It was part of the six car series GATX 1800-1805.
(Paul C. Winters)

GATX 8338
ICC111A100W2
◀ At first glance a plain car, but closer inspection shows an unusual type of gutter applied. It was built in November 1965 (others were built the prior month) and part of a 25-car set numbered 8332-8356. The cars were described as being in acid service. *(Craig Bossler)*

GATX 10736
DOT111A100W5
◀ A car in muriatic acid service when put on film in July 1984 at Maumee, Michigan. The build date on this 20,703 gallon car is September 1969 and it was one of six lined cars in the set 10732-10737.
(Morning Sun Books collection)

GATX 19531
DOT112A340W
▲ A photo taken on February 11, 1973. December 1969 is when this car was constructed and it was assigned to the 23-car series 19521-19543. It was leased to a Houston based wholesaler of liquid natural gas, founded in 1968. *(Dave Nelson, David Hickcox collection)*

GATX 20805
DOT103W
▶ Another stub sill car with a dome. The car, found in Wichita, Kansas, was fabricated in February 1974 and has a sparger system installed. Of 12,765 gallon capacity it was part of the eight car set 20801-20808.
(Jim Kinkaid)

GATX 26296
DOT111A100W1
▼ That October 1974 date applied at Sharon (Pennsylvania, General American's plant) was when this car was manufactured (it is believed that more cars were assembled in September and November). The car was first listed in the April 1975 equipment register when it was annotated as part of the 26290-26299 group. It was spotted in Decatur, Alabama in January 1976 and was a 20,538 gallon car fitted with heater coils.
(Bernie Wooller)

GATX 38581
DOT112A400W
▲ Built in February 1969 this car is in propylene service. A 33,708 gallon car, it was recorded on film in April of that year. It was part of a 33-car set numbered 38556-38588. *(Paul C. Winters)*

GATX 38751
DOT111A100W5
▼ At Painesville, Ohio on August 7, 1971. Assembled in November 1965, this rubber lined car received this paint job at Sharon in July 1968. The register has the car within the 38746-38755 number series.
(Dave McKay, Morning Sun Books collection)

GATX 40951
DOT112A340W
◀ A car found at Cozad, Nebraska on May 25, 1973. As may be seen, it was furnished by General American in October 1970. It was assigned to the GATX 40800-40951 number series which comprised various lessee's and were constructed from September 1970 into early 1971. Compare the paint scheme to the one in Volume One, page 34.
(Jim Kinkaid collection)

GATX 50468
DOT111A60W2
▲ A three-car TankTrain at Dearborn, Michigan in January 1977 for sulfuric acid service. The three cars here (GATX 50466-50468) were assembled in April 1976. TankTrain was a novel approach by General American to reduce loading/unloading time by interconnecting the cars. As can be seen, the concept worked well for short car sets too. *(Emery Gulash, Morning Sun Books collection)*

GATX 83545
◄ An 8,266 gallon car, it was photographed in March 1972. Constructed at Sharon in October 1969 it was part of the 83540-83549 group. Production may have also extended into November. *(William Rosenberg)*

GATX 85086
DOT111A100W5
▼ At Belt Line Junction in Pennsylvania on June 14, 1978. Fabricated in August 1964 it was a 20,427 gallon car carrying muriatic acid. At the time of the photo the car was one of two, the other being car 85087. *(Craig Bossler)*

GATX 87272
DOT111A100W1
▲ Seen at Wichita, Kansas in May 1988, the car had been assembled in October 1969. It was a part of the set 87249-87299. Sun Oil Company, "Sunoco," was one of the largest independent refiners of petroleum products in the United States. *(Jim Kinkaid)*

GATX 88445
▶ A photo taken in March 1974. It is thought that the car had been assembled in February 1967. The first register listing was in July 1967 as one of 22 cars in the series 88435-88457. The company leasing this car specialized in plasticizers and pesticides and was based in Rosemont, Illinois. *(Bernie Wooller)*

GATX 89927
ICC112A340W
▶ Birmingham, Alabama was where this car was at in November 1966. Note the mud shields above the wheels. The car was part of the GATX set 89902-89936. It is thought that the cars had been assembled in February and March of that year. American Cyanamid, formed in 1907, was a chemical manufacturer.
(Jim Thorington)

GATX 91249
ICC112A340W
▼ Another car with mud guards also found in Birmingham, Alabama, but in July 1968 (Wanda Petroleum having been just acquired by Ashland Oil two months earlier). Constructed in November 1967 it is believed that this particular car was part of the 91212-91256 group. *(Jim Thorington)*

GATX 92358
DOT111A100W1
▲ In December 1971, also at Birmingham, Alabama. The car was one of 75 in the set 92349-92388 which had been assembled in November and December 1968. Apparently all were leased to the Dow Company.
(Jim Thorington)

GATX 98243
◄ A car placed on film in May 1971. It was part of the 98243-98249 number series. These cars are thought to have been built in November 1968.
(William Rosenberg)

GCX 420028
DOT114A400W

◀ The car was manufactured in June 1971 and is a 20,842 gallon car. It was one of 14 such cars. Originally owner Allied Chemical had the cars listed in the register as 42016-42030. The clerical error was not discovered when the GCX-marked cars were transferred to the Alltank Equipment Company in October 1972. It was not until January 1974 that the correct numbers and series was figured out, which was 420016-420029. *(William Rosenberg)*

GROX 5018

◀ At Reading, Pennsylvania on May 10, 1992 was this car which had been constructed by General American Transportation Company. Possibly built in October 1969, the 23,710 gallon car did not appear under these reporting marks (first used by Chrysler Rail in April 1989) until January 1990. Thus its series, 5000-5043, was acquired used. *(Craig Bossler)*

GRYX 2009
DOT111A100W1

▼ Owned by the John H. Grace Company, the stenciling on its side used to state "National By-Products Inc" before all the crud covered much of it. The coil equipped 20,460 gallon car was in Wichita, Kansas in August 1989. Assembled by American Car and Foundry in November 1964 it was part of the 2000-2025 set. *(Jim Kinkaid)*

HPLX 451037
DOT111A100W1
▲ Here is a car manufactured by Union Tank Car in May 1969 for the Humble Pipeline Company (later a part of the Exxon Corporation). It has the extra air reservoir because it is a 125-ton capacity car. Rated at 30,000 gallons this example was part of the set 451021-451040. It was at Coffeyville, Kansas in September 1989. *(Jim Kinkaid)*

KCCX 2650
DOT111A100W2
▶ A car in sulfuric acid service that was five months old when caught on film in November 1987. It was built by Union Tank Car in May 1987 for owner Kennecott Copper under lot 6430. There were 120 of these in the series 2600-2719.
(Emery Gulash, Morning Sun Books collection)

MCVX 79116
DOT111A100W1
▼ Baltimore, Maryland in June 2002. It would appear to be a training car since it is placarded "do not load." The car, likely built by North American Car, was never listed in the equipment register. *(Jim Rogers)*

MONX 211
ICC111A100W5
◀ A brand new car at Huntsville, Alabama in July 1969. It was fabricated only the previous month by General American. Rubber lined and of 20,540 gallon capacity, it was part of the 200-222 set owned by the Monsanto Company. *(Emery Gulash, Morning Sun Books collection)*

NATX 17148
ICC112A400W
▲ It is June 1968 in this picture. The subject was assembled by North American in November 1962 (although the tank itself was manufactured by the J.B. Beaird Company) and assigned into the 17100-17199 number group. *(Paul C. Winters)*

NATX 22758
ICC111A100W1
▶ April 1972 was when this 20,479 gallon car was photographed. It has April 1968 manufacture markings from the builder's Texarkana, Texas plant. There were ultimately 90 of these cars placed into the 22703-22792 set. *(William Rosenberg)*

NATX 23127
DOT111A100W1
▲ At one point in time it was stenciled "VOP" to the left of the ladder. To the right: "Vegetable Oil Products Company, Inc." was on the top line and below that: "World's Largest Independent Producer of Fats and Oils." The coil equipped car, spotted in April 1972, was constructed in November 1968 by North American Car at Texarkana and belonged to the group 23100-23166. *(Paul C. Winters)*

NATX 25080
ICC111A100W4
▼ An illustration taken in May 1979 of a car assembled at Texarkana, Texas by North American Car in July 1965. Equipped with heater coils it was part of the 36-car set numbered 25056-25091. *(Jim Thorington)*

NATX 25645
DOT111A100W1
▶ On the move at Kansas City, Missouri on March 25, 1992. This car was fabricated at Texarkana in September 1972 and was one of 15 cars numbered as 25632-25646.
(Jim Kinkaid)

NATX 29649
DOT111A100W1
▲ I've no idea if this was an early attempt at a bicentennial paint scheme or not, but the car was completed at Texarkana in December 1973 and was found at Atlanta, Georgia on May 23, 1987. It was leased to the Grain Processing Company and had a tank capacity of 30,105 gallons. There were 100 cars in its assigned series 29600-29699, apparently leased to various companies. *(Jim Kinkaid collection)*

NATX 34647
DOT112S340W
▲ Here is a car sitting at Little Rock Arkansas on February 7, 1984. It has been modified with head shields, which are clearly visible in this illustration. The car was part of an extensive group which ultimately numbered 1,289 cars (series 33800-35088) probably built under four (or more) orders between late 1965 and mid-1973. It was one of 534 first reported in early 1970. *(Al Lanier, Jim Kinkaid collection)*

OILX 30675
DOT111A100W1
▼ Assembled by Richmond Tank Car under lot 3443, the car sits at Nampa, Idaho in August 1982. The 100 cars in this set (OILX 30600-30699, leased through North American Car) were manufactured in May and June 1982. This was a 30,500 gallon car. *(Al Lanier, Jim Kinkaid collection)*

POTX 139
ICC112A340W
▶ This Pargas Incorporated car was built in March 1966 and part of the 40-car series 101-140. It was at Trenton, New Jersey in August 1967 and had a capacity of 33,500 gallons.
(William Rosenberg)

PROX 16072
DOT111A100W2
▼ A Procor Incorporated car constructed by Union Tank Car at East Chicago in June 1995; an item of note is the French text on the upper right hand end of the tank. The car, put on film in July 1998, was never listed separately in the equipment register but is thought to have belonged to the set 16059-16315, assembled under lot 5107.
(Emery Gulash, Morning Sun Books collection)

PROX 50204
DOT111A100W1
▲ We're looking at a picture taken on May 21, 1992 at Richmond, Virginia on the Richmond, Fredericksburg & Potomic Railroad. Assembled by Union Tank Car in April 1991, it would become part of the 70-car set 50200-50269. *(Gib Albach)*

QFFX 25161
DOT112S500W1
◀ A car owned by the Mexican company Quimica Fluor, S.A. de C.V. sits in Memphis, Tennessee on June 28, 2001. This was a 22,263 gallon capacity car and as can be seen by the builder's logo, it was constructed by American Car and Foundry. There were 23 of these cars in the series 25155-25177 which were furnished in April through June 2000. *(Al Lanier, Jim Kinkaid collection)*

RAIX 6563
DOT111A100W1
▲ United States Leasing International owned this one, found sitting at North Little Rock, Arkansas on March 13, 1986. Leased to Union Carbide it was part of the set 6500-6576. The 20,000 gallon car had been assembled in January through March 1976 by Union Tank Car's East Chicago plant.
(Al Lanier, Jim Kinkaid collection)

RCRX 1476
▼ Here's a car in hydrochloric acid moving through Tullahoma, Tennessee on September 23, 2001. Constructed in June 1998 (additional cars were also built that May) it became part of the 50-car group numbered 1445-1494. They had been assembled by Trinity Industry's Longview, Texas plant under file 97113. The 22,900 gallon car was owned by Reagent Chemical and Research, Incorporated based in Houston, Texas. *(Lee Yoder)*

RELX 2223
DOT111A100W1
▶ It must be nice to actually own your own tank car! Photographed in May 1972 is a car that was part of the 2201-2299 set (the cars may have had odd numbers only). Coil equipped it was assembled in December 1971. These reporting marks were used by Relco Leasing, which became Relco Tank Line circa 1972. *(William Rosenberg)*

RELX 2411
DOT111A100W1
▼ At Wichita, Kansas in May 1988 is a car furnished by American Car and Foundry which was of 20,000 gallon capacity and fitted with coils. It was part of the 2400-2499 series, which were known to have been constructed in May and July 1972. *(Jim Kinkaid)*

RTMX 2064
DOT111A100W5
▼ The Richmond Leasing Company operated these cars. The one shown here was at Baltimore, Maryland in October 1987. It was supplied by Richmond Tank Car in March 1972 as a 20,000 gallon car. The series was 2050-2099. *(Jim Rogers)*

RTMX 2619
DOT111A100W3
▲ Photographed in October 1979 this was a relatively plain car in styrene service, which is of course what many a model railroad car is made of. The car, built in April 1973, was part of the 2600-2639 number set. These 26,000 gallon cars were supplied by Richmond Tank Car under lot 2145 and were delivered from March through May 1973. *(Rail Data Service, Jim Kinkaid collection)*

TLDX 114057
DOT111A100W1
◀ Pullman-Standard acquired two of these cars under their lot TLR-1023 for 54% phosphoric acid service. Richmond Tank Car built them in May 1970, to be numbered TLDX 114057-114058. The cars were bottom load only and carried 14,500 gallons of the chemical. Here is a builder's view of one. The "TLD" within the reporting mark stood for Transport Leasing Division of Pullman-Standard.
(Pullman-Standard photo, Jim Kinkaid collection)

TLDX 120021
DOT111A100W1
▼ Another builder's photo taken by Pullman-Standard. The car was part of the 16-car series 120010-120025 which were built as 20,800 gallon capacity cars. They did have heater coils. The set was delivered by Richmond Tank Car in November 1969.
(Pullman-Standard photo, Jim Kinkaid collection)

TVAX 2008
DOT111A100W1
▲ An almost new car sits in Muscle Shoals, Alabama on April 29, 1981. It was built just two months earlier by American Car and Foundry. The 21,000 gallon car was listed as part of the 6-car set 2008-2013. *(Al Lanier, Jim Kinkaid collection)*

TVAX 2087
DOT111A100W1
▼ A car that was part of the 2019-2097 set as listed in the equipment register. There may have been multiple orders involved and cars were known to have been constructed from August 1981 through January 1982. *(Lee Yoder)*

UCLX 30144
DOT111A100W5
▶ This one, owned by Vulcan Materials' Chemical Division, sits at Wichita, Kansas on March 15, 1992. It was of 20,536 gallon capacity and assembled by Trinity Industries at their Oklahoma City, Oklahoma plant in September 1991. The car belonged to the set 30135-30154. *(Jim Kinkaid)*

UTLX 13152
DOT111A100W2

▶ Although in plain black, this 13,853 gallon car is unusual due to the length of the end platforms. (It may have been configured this way in an attempt to solve car tracking issues.) It was placed on film at Tullahoma, Tennessee on April 26, 1981 and was in sulfuric acid service. Assembled in March 1971 it was part of the UTLX series 13140-13161.

(Lee Yoder)

UTLX 13304
ICC111A100W2

▶ Having been built in August 1965 it was found sitting the following October. The car was rubber lined for phosphoric acid service and Union Tank Car documents show that it carried 14,612 gallons.

(Bernie Wooller)

UTLX 13339
ICC111A60W1

▶ Also constructed in 1965 (but in October) this example was caught on film the following January. It too was rubber lined. This particular car was of 14,654 gallon capacity. The lessee was based in Streater, Illinois and was one of America's largest corporations devoted to fertilizer manufacture.

(Bernie Wooller)

UTLX 13457
ICC111A100W2
▲ This one was manufactured in July 1966 and spotted in Decatur, Alabama in January 1976. Used in phosphoric acid service and PVC lined, it was of 14,806 gallon capacity. The equipment register has this car within the 13428-13483 group. *(Bernie Wooller)*

UTLX 41993
AAR211A100W1
▼ Assembled in June 1970 this car was found in July 1973 and hauled 20,528 gallons. It was part of the 7-car set 41993-41999. Of note is the fitting near the reporting mark, a sampling spigot perhaps? *(Paul C. Winters)*

UTLX 48452
DOT111A100W1
◀ Fabricated by Union Tank Car in February 1970 it was placed on film in September 1972. Its tank held 20,545 gallons. While shown as a single car in the register, there were most likely other similar cars under lease to Sun Oil with scattered car numbers.
(Paul C. Winters)

UTLX 48888
DOT111A100W1
▲ Another Heniz car found in July 1973. It too has that fitting next to the reporting marks. The car was reported to have a capacity of 19,438 gallons. It's possible that the car belonged to the UTLX series 48886-48899. *(Paul C. Winters)*

UTLX 49989
DOT111A100W1
▲ In May 1964 the photographer caught this car. The entire car was manufactured in September 1959 although the tank had been furnished by Graver Tank and Manufacturing the previous month (and so stenciled). It is thought that the car belonged to the series 49985-49998. *(Paul C. Winters)*

UTLX 50000
DOT111A100W1
▲ A nice looking car photographed in October 1966. One of two such cars (the other being 49999) it was also assembled in September 1958, the Graver-furnished tank having been constructed that July. *(Paul C. Winters)*

UTLX 58250
DOT111A100W1
▲ Although in plain black and yellow, the combination of funnel flow design and reinforcement rings makes this a rather unusual design. It was at Rochester, Pennsylvania in June 1990. While it shows an October 1974 manufacture date, the car has been recently repainted: the chalk line is still visible! Fitted with heater coils it was of 26,709 gallon capacity. The car belonged to the group 58237-58266, furnished under lot 3458.
(Jim Kinkaid)

UTLX 58709
DOT111A100W1
▶ Union Tank Car developed the Funnel Flow tank car design, wherein the tank shell was built in such a way that both ends sloped towards the outlet. A look at one is illustrated in this shot taken in December 1971. It's hard to pinpoint the exact series that this 20,473 gallon car belongs to, but data suggests the set 58698-58712.
(Bernie Wooller)

UTLX 58828
DOT111A100W1
▲ A newly repainted Funnel Flow at Kansas City, Missouri in 1995. It is a 20,479 gallon car under lease to 4-G Processing and was assembled in September 1971. More cars known to be like this one were UTLX 58779-58837, some built in August. *(Jim Kinkaid)*

UTLX 59523
DOT111A100W
▶ The car was manufactured in February 1967 and had a capacity of 20,623 gallons. Documents show that the car was part of the UTLX set 59518-59530.
(Bernie Wooller)

UTLX 80690
▼ Built in October 1963 and of 33,758 gallon capacity, it was photographed in July 1971. The Getty logo belonged to the famous J. Paul Getty oilman of Tulsa, Oklahoma.
(William Rosenberg)

UTLX 83053
▼ A view of a car that seems about as long as any hi-cube box car found sitting at Tullahoma, Tennessee in June 1963. That long tank held 30,149 gallons. The photographer's notes say that it was carrying Spensol, an industrial coating product mentioned in Volume One. *(Lee Yoder)*

UTLX 83078
▲ Another car thought to be from the same series as 83053. It was found in Reading, Pennsylvania on July 5, 1988 in anhydrous ammonia service. The catwalk supports and bracing are noteworthy on this car. This was a 30,170 gallon capacity car. Documents suggest that both cars were part of a set numbered 83013-83080, though this has not been confirmed. *(Craig Bossler)*

UTLX 88166
ICC112A340W
▼ This photograph was taken in March 1968. It is of a 33,729 gallon car assembled in March 1967. At the time of the picture it was described as being part of the UTLX group 88144-88259. *(Paul C. Winters)*

UTLX 99398
▼ A fascinating scene taken at a railroad equipment display in San Francisco, California on June 11, 1960. The car appears to have an April 1960 build date and records indicate that it carried 29,940 gallons. It was one of two such cars, the other one was numbered 99399.
(David Graeff, Morning Sun Books collection)

UTLX 200645
DOT111A60W7
▼ Spotted sitting in Kansas City, Missouri in September 1995, the car is only five months old. The car was under lease to the Degussa Corporation, the North American arm of the German-based specialty chemical consortium. It was part of the 200626-200650 number series, which were assembled in February through April 1995. *(Jim Kinkaid)*

UTLX 201631
DOT111A100W1
◀ A car found at San Bernardino, California on February 6, 1993. It has a capacity of 30,128 gallons and was part of the number set 201603-201632, leased to Union Carbide. These cars were built in February and March 1992 at the East Chicago plant.
(Jim Kinkaid)

UTLX 201676
DOT111A100W1
◀ In the same Santa Fe train was this car, quite similar to the one above. It however is a 23,461 gallon car and was fabricated in September 1992. It was a part of the group 201664-201697. Also from Union Tank's East Chicago shops the cars were delivered in September 1992 under lot 6897. They too were under lease to Union Carbide.
(Jim Kinkaid)

UTLX 970024
DOT112S500W

◀ Rated at 24,048 gallon capacity the car was constructed in March 1990: that 500 psi tank rating is a bit unusual. It was one of 45 such cars numbered as 970000-970044. More cars were also manufactured that February and all may have been under lease to the DuPont Corporation. The illustration was taken in Memphis, Tennessee on June 28, 2001. The commodity carried is a source of the element fluorine, used in such diverse applications as Teflon® and pharmaceuticals.

(Al Lanier, Jim Kinkaid collection)

Single Compartment Insulated Cars

ACFX 71466
DOT111A100W1

▲ This photo was taken in September 1988. Assembled in August 1986 under builder's lot 71-12339 it was one of 15 in the set 71463-71477.

(Emery Gulash, Morning Sun Books collection)

ACFX 75603
DOT111A100W1

▶ At Fulton, Kentucky in July 1983. It was part of the 30-car group numbered 75594-75623 built in May 1982 under lot 71-12302. The car had a capacity of 13,354 gallons.

(Al Lanier, Jim Kinkaid collection)

ACFX 75735
DOT111A100W1
◀ Another Georgia Kaolin car at Kansas City, Missouri on April 16, 1994. This one was one of six numbered 75712-75756 that were originally painted for Anglo American Clay. The cars were built in the first half of 1983 under ACF lot 71-12305. *(Jim Kinkaid)*

ACFX 76623
DOT111A100W1
◀ Here is a slightly different Georgia Kaolin paint scheme found in Birmingham, Alabama in May 1985. The car was part of the 50-car set ACFX 76610-76659, assembled in August and September 1984. *(Jim Thorington)*

ACFX 76965
DOT111A100W1
▼ A car under lease to Union Carbide put on film in November 1987. Fabricated in June 1985 (more cars were assembled in July) it was one of 23 such cars in the 76954-76976 set.

(Al Lanier, Jim Kinkaid collection)

ACFX 77624
DOT111A100W1
▲ Another scheme photographed in June 1976. It was passing through Newark, New Jersey at WA-5 tower when caught on film and was built in May 1976. The car was one of 20 in the group 77614-77633 and it was of 13,423 gallon capacity. The set was assembled under builder's lot 71-12258 with additional cars supplied that February.
(William Rosenberg)

ACFX 77645
DOT111A100W5
▼ A car under ADM (Archer-Daniel-Midland) lease while at Memphis, Tennessee on February 11, 1981. Constructed in April 1976 with heater coils it was part of the 15-car set 77634-77648.
(Al Lanier, Jim Kinkaid collection)

ACFX 77744
AAR211A100W1
▼ Here is a car found in Ashdown, Arkansas on February 19, 1986. Built in June 1976 it was part of the six car series 77742-77747, furnished under lot 71-12260. *(Al Lanier, Jim Kinkaid collection)*

ACFX 78275
DOT111A100W1
◀ Another Anglo-American Clay car but with a different reporting mark style. Note that even though the car appears similar to the one previous, the tank is built to a DOT specification. It was at Lakeland, Florida in January 1978, having been assembled in May 1977 originally for Thiele Kaolin. The car was part of a 35-car group numbered 78250-78284.
(Emery Gulash, Morning Sun Books collection)

ACFX 78513
DOT111A100W1
▲ Ten of these cars were fabricated by ACF in March 1978 as ACFX 78506-78515 from lot 71-15296. Kansas City, Missouri was where this one was photographed on March 26, 1992. Terra Industries, the lessee of the cars, was a company formed in 1964 as Terra Chemicals International to manufacture nitrogen products. It was based in Sergeant Bluff, Iowa. *(Jim Kinkaid)*

ACFX 78559
DOT111A100W1
◀ Another Georgia Kaolin car found in July 1979. It was constructed at ACF in April 1978 as one of 20 such cars placed into the 78540-78559 number group. The cars were built under lot 71-12279.
(William Rosenberg)

ACFX 78638
DOT111A60W1
▲ A car at Birmingham, Alabama in March 1980. There were 20 of these assembled in June and July 1978 under lot 71-12280. The set was given the numbers 78624-78643. *(Jim Thorington)*

ACFX 78871
DOT111A60W1
▶ Another one at Kansas City, Missouri on April 17, 1994. It was part of a 10-car batch numbered as 78866-78875 from lot 71-12281 and had a capacity of 13,633 gallons. *(Jim Kinkaid)*

ACFX 79253
DOT111A100W1
▼ On that same date and in the same Burlington Northern train was this car. It was assembled in September 1980 and the equipment register showed it to be part of the 79242-79271 group. More cars from this set were assembled in October and all derived from lot 71-12294. *(Jim Kinkaid)*

ACFX 83675

◀ On January 20, 1972 this car was photographed at Othello, Washington. It was built in August 1971 and was of 21,019 gallon capacity. While the equipment register shows it listed as part of the group 83656-83689, it is thought that Honeymead only leased cars 83671-83685. *(Wade Stevenson)*

ACFX 87146

AAR211A100W1

▼ Yet another car in kaolin slurry service, in this case from a company out of Gordon, Georgia. The car, of 13,632 gallon size, was at Memphis, Tennessee in December 1981. It was part of the set 87142-87155, manufactured in July and August 1974.

(Al Lanier, Jim Kinkaid collection)

ACFX 87970

AAR211A100W1

▲ At Emery Gap, Tennessee on October 16, 1982. This 13,630 gallon car was part of a set of clay slurry cars constructed in October and November 1975. It was one of 12 cars assigned numbers 87968-87979, although various kaolin producers received cars out of this small set. *(Al Lanier, Jim Kinkaid collection)*

ACFX 88818
ICC111A100W1
▶ A picture taken in September 1971. Built in October 1968 it was one of 50 cars in the set 88801-88850 and it does have heater coils.
(Paul C. Winters)

ACFX 89030
ICC111A100W1
▼ The chemical carried by this car (constructed in November 1968 and photographed the following April) is normally distilled from coal tar and is used primarily as an intermediate chemical (meaning it is utilized to make other chemicals), although it is also used in ordinary moth balls! Some of the 14 coil equipped cars in this group (89022-89035) were also fabricated in December. *(Paul C. Winters)*

ACFX 89970
DOT111A100W1
◀ A car found in Baltimore, Maryland in October 1994. The 23,570 gallon car had been constructed in September 1971. As such it was one of four in the series 89969-89972 originally painted for the Uniroyal company.
(Jim Rogers)

ACFX 94925
AAR211A100W1
▼ In July 1993 the car shown here was assembled as part of the set 94919-94939. They were leased to the SCM Corporation (the corporate successor to Smith-Corona Incorporated and Marchant Calculator companies) and used in their coatings and resin manufacturing operations. The picture is dated June 1998. *(Emery Gulash, Morning Sun Books collection)*

ACFX 95406
AAR211A100W1
◀ A clay slurry car photographed in October 1997. It was assembled in September 1994 and was one of 30 in the series 95388-95417. *(Emery Gulash, Morning Sun Books collection)*

ADMX 15110
DOT111A100W1
▼ At Tullahoma, Tennessee in March 2006 was this General American-built car. The 550-car group it was a part of (15001-15550) was assembled in December 1980 through July 1981, possibly under builder's lot 10697. These were 17,600 gallon cars. *(Lee Yoder)*

ADMX 18020
AAR203W
▲ Another ADM-owned car found at Tullahoma, Tennessee, but on September 9, 1998. Resembling an oversized prune, the 18,228 gallon car was originally part of a set of cars constructed by American Car and Foundry between 1969 and 1974 for corn syrup. In 1985 and 1986, some were rebuilt to this Uni-temp design. At the time of the photo there were still 17 in revenue service in the series 18005-18025. *(Lee Yoder)*

ADMX 19061
AAR211A100W1
◀ Assembled in May 1995 by Trinity at their Longview, Texas plant the car was found at Baltimore, Maryland in January 2004. It was described in the equipment register as part of the 18-car set numbered 19045-19062.
(Jim Rogers)

AMPX 17566
DOT111A100W3
▼ Leased through North American Car, it was one of 100 in the series 17500-17599. Assembled by Union Tank Car under lots 3101N and 3102N in January 1977, they were of 17,500 gallon capacity. Our example was found at Memphis, Tennessee on April 26, 1982. The reporting marks show that the car was under lease to the American Maize Products Company, incorporated in 1906 and who marketed their Tru-Sweet brand of high fructose corn syrup. *(Al Lanier, Jim Kinkaid collection)*

AMRX 122
DOT111A100W1
▼ Ameropan Oil, based in Illinois, was the owner of this car. The company was a producer of asphalt and operated 20 of these cars in that service, being manufactured in August and November 1989 by Trinity Industries under file 8838E. For some reason the series was numbered 101-128. Car 122 sits at Reading, Pennsylvania on May 3, 1991. *(Craig Bossler)*

ASRX 1685
DOT111A60W1
▼ An American Sugar Company (better known as Domino Sugar) car at Baltimore, Maryland in October 1982. It was manufactured by General American Transportation in October 1968 and was of 16,000 gallon capacity. Originally fabricated to an ICC111A100W1 specification, it was part of the 1651-1685 set. *(Jim Rogers)*

ASRX 1700
DOT111A60W1
◀ From the series 1700-1714 manufactured in May 1970, a group not listed in the equipment register until October of that year. (By the following January, the company had been renamed Amstar.) Built by ACF, the 17,652 gallon car sits at Council Bluffs, Iowa on September 6, 1988.
(Jerry Bosanek, Jim Kinkaid collection)

ASRX 1716

DOT111A100W3

▲ Another liquid sugar car found in Baltimore in November 1987. Assembled in December 1969 by Union Tank Car, it was of 16,700 gallon capacity. It belongs to the small series 1715-1717. *(Jim Rogers)*

ASRX 1729 & 3000

DOT111A100W1

▶ ▼ More liquid sugar cars. At the mid-position is a car assembled in November 1971 by Union Tank Car. With a marked capacity of 17,514 gallons it was listed as part of the 1724-1749 set, although the equipment register showed the series to be of 17,120 gallon size. It was photographed in April 1985. Below it is another ASRX car at Baltimore, having been spotted there in February 2002. The car was manufactured in September 1997 at Trinity's Oklahoma City, Oklahoma plant. As such it was one of 25 in the set 3000-3024, assembled under builder's file 9749.

(middle, Al Lanier, Jim Kinkaid collection; bottom, Jim Rogers)

CCLX 1517
AAR203W
▼ Besides the numerous tank cars with full length center sills shown in Volume One, Crystal Car Line also utilized stub sill design cars. From a 44-car set is this one found at Altoona, Pennsylvania in June 1990. Apparently built by General American in February 1976 as a 16,000 gallon car, it was assigned to the number series 1501-1544. *(Jim Kinkaid)*

CCLX 3121
DOT111A60W1
◀ A different version of the paint scheme taken in June 1973. As may be seen it was built by General American in July 1967. The 8,000 gallon coil equipped car was part of a 117-car order that resulted in car numbers 3097-3213.

(Paul C. Winters)

CELX 3206
DOT111A100W1
▼ A 21,000 gallon Celtran car sits at North Little Rock, Arkansas on December 7, 1983. Constructed in December 1969 by Richmond Tank Car it was used primarily to carry acetic anhydride. More of these cars were assembled in January 1970 which resulted in the series 3200-3282.

(Al Lanier, Jim Kinkaid collection)

CELX 60000
DOT111A100W6
▲ Originally carrying UTLX markings the car was transferred to these reporting marks about January 1997. Built by General American in August 1978 as UTLX 60000-60038, some 17 cars moved to Celtran, keeping their original numbers. This 20,603 gallon car was at Tullahoma, Tennessee on November 21, 1998. *(Lee Yoder)*

CGTX 15832
DOT111A100W1
▲ The Canadian firm Hawker Siddeley fabricated this car in June 1980. It was part of the Canadian General Transit Company set 15823-15835, having been assembled under company order 5799. This view was taken in October 1993. *(Emery Gulash, Morning Sun Books collection)*

CGTX 64207
DOT112J340W
▲ By the time this car was spotted at Nampa, Idaho in March 1990 the car's owner had changed its name to CGTX Inc., thus the plain CGTX logo. Part of the CGTX series 64197-64221, it was assembled at Hawker Siddeley in August 1970 under company order 5400. *(Jim Kinkaid)*

CITX 31552
DOT112J340W
◀ The reporting mark CITX was owned by PLM Canada, who's first listing was January 1987. By April of that year this car, photographed at Slidell, Louisiana on April 25, 1995, was one of four tabulated in the equipment register as part the group 31550-31566. Constructed in May 1967, likely by North American, the cars may in fact be ex-NATX, keeping their same numbers when moved over to CITX markings.
(Lee Yoder)

CRGX 6102
▼ Cargill owned this car, one of 36 in the set numbered 6089-6124. Assembled by Trinity in February 1995 it was found in Tullahoma, Tennessee in March 2006. The FRA-mandated conspicuity stripes are hard to miss, which is the point of the exercise! *(Lee Yoder)*

CRGX 6281
DOT111A100W3
◀ In earlier days with full paint. Put on film in March 1998 it was fabricated by Trinity's Oklahoma City plant in October 1994. Equipped with heater coils it was part of the 6125-6302 set built under file 9413 in September through November 1994. To that were added cars 6303-6486, constructed between August 1994 and January 1995 under file 9413W.
(Emery Gulash, Morning Sun Books collection)

CSOX 23644
DOT111A100W1
▼ A Cities Service Company car at Wichita, Kansas on April 28, 1991. It was constructed nearly ten years earlier, in May 1981. The 96 cars in the group 23600-23695 were also built at Trinity Industries in April 1981 and were of 23,500 gallon size. *(Jim Kinkaid)*

DAX 7098
DOT111A100W1
▶ While lacking the snappy Diamond Shamrock paint scheme seen in Volume One, the car is still nicely painted grey and black as seen sitting at Kansas City, Missouri on March 25, 1992. It was built in January 1981 as part of the 32-car series 7068-7099. The cars were probably acquired used since they did not appear in the equipment register until April 1987. *(Jim Kinkaid)*

DNAX 123026
DOT111A100W1
▼ A Dana Railcare car at Ft. Meade Junction in Maryland in December 2005. It had been built in October 1995 by Trinity Industries under file 9584. There were 32 of these cars, which were numbered 123000-123031. *(Jim Rogers)*

DUPX 8603
DOT111A100W1
◀ Ti-Pure is DuPont's brand name for their line of titanium dioxide products. As a supplier of 25% of the world's production of this handy stuff (used in such things as paint, plastics and paper manufacturing), the company ships a *lot* of it. Here is one car in this service at Decatur, Alabama in January 1976.
(Bernie Wooller)

DUPX 8604
DOT111A100W1
▼ Another car from this same set. Both cars were part of the 10-car series 8600-8609 constructed by American Car and Foundry in late 1971. This one was at Birdsboro, Pennsylvania on September 16, 1973, having been assembled in October 1971. The cars were of approximately 11,280 gallon capacity.
(Bernie Wooller)

DUPX 8618
AAR211A100W1
▲ Although listed in the equipment registers with the above Ti-Pure cars just illustrated, note that this one is of a funnel flow design. It was found in Altoona, Pennsylvania in June 1990. *(Jim Kinkaid)*

DUPX 8619
AAR211A100W1
▶ Mate to 8618 (shown on the previous page) is this 11,236 gallon car seen passing through Tullahoma, Tennessee in March 2002. Both were part of the 8610-8631 series, probably assembled in late 1972.
(Lee Yoder)

DUPX 8641 & 8671
AAR211A100W1
DOT111A100W
▼▼ More titanium dioxide cars. At middle is one at Altoona in September 1989. It was part of the 48-car set 8632-8679, which had been constructed from late 1975 through January 1976. This car has been converted from its original DOT specification. Below, from the same group is a car found at Mojave, California on May 3, 1986. *(middle, Jim Kinkaid; bottom, Pat Holden, Jim Kinkaid collection)*

DUPX 20034
AAR206W
▲ An interesting find photographed in April 1969. Probably built by North American Car around late 1967, it was of a "tank-within-a-tank" construction. One of 10 in the group 20030-20039, the car featured two structural tanks separated by heavy insulation to afford maximum protection to heat sensitive ladings without having to resort to vacuum systems. The car was of 23,400 gallon capacity. *(Paul C. Winters)*

DUPX 29530
DOT111A100W4
▲ Moving down grade at Altoona, Pennsylvania in June 1990. The peculiar cutout along the side of the tank is for clearance purposes, so that the car will remain within Plate "C" limits. The product stenciled on the side is an organic chemical used to make various polymers such as DuPont's Nylon 66® and polyurethanes. The car is part of a 72-car set numbered 29500-29574, built around late 1966 or early 1967. *(Jim Kinkaid)*

GATX 12729
DOT111A100W1
◄ A car at St. Louis, Missouri on April 26, 1981. It is part of the set 12721-12730 fabricated in November 1977 by General American.
(Michael Wise, Jim Kinkaid collection)

GATX 13215
DOT111A60W1
▲ In kaolin slurry service while at Birmingham, Alabama in April 1980 is a car that belonged to the 13211-13220 series, although there were other lessees involved too.
(Jim Thorington)

GATX 16155
DOT111A100W1
▶ Here is a car in sulphur service while at Tullahoma, Tennessee in February 2004. It was part of the set 16151-16180 furnished by Trinity Industries in January 2000. The 13,600 gallon cars were supplied under file 9974. *(Lee Yoder)*

GATX 16317
DOT111A60W1
▼ Another kaolin slurry car at Memphis, Tennessee in March 1981. It was a part of the 16301-16320 group constructed in May and June 1978.
(Al Lanier, Jim Kinkaid collection)

GATX 18132

◀ Another car which was found in Kansas City, Missouri on April 17, 1994. Apparently built in March 1979 it was a 14,017 gallon car and assigned into the series 18131-18150. *(Jim Kinkaid)*

GATX 19849

AAR211A100W1

▼ Yet another kaolin slurry car. This one was photographed in October 1997. It was fabricated in December 1978. Information suggests that it was part of the set 19817-19865, built by General American between September and December 1978. *(Emery Gulash, Morning Sun Books collection)*

GATX 20044

DOT111A60W1

▼ The Georgia Kaolin car here was found at Memphis, Tennessee on September 2, 1981. As such it was one of 25 in the series 20026-20050 assembled in September 1979 under lot 10426. *(Al Lanier, Jim Kinkaid collection)*

GATX 22344
DOT111A100W1
▼ Yet another car in kaolin service. This photograph was taken in September 1985. The car was assembled by Trinity Industries in May 1985 under file 8440. It was part of the set 22331-22355. *(Emery Gulash, Morning Sun Books collection)*

GATX 25135
DOT111A100W1
▼ This car likely belongs to the series 25101-25154, built in February and March 1970, originally leased to the Clinton Corn Processing Company. Furnished in March 1970 its photograph was taken on June 7, 1992 at North Lake, Illinois. *(Gib Albach)*

GATX 28351
DOT111A100W1
▲ At Wichita, Kansas in May 1991 is a car that was one of four leased to Unichema International, who was a major supplier of fatty acids and oleo-chemicals. It was assembled in March 1987 by Trinity Industries and part of the number set 28351-28354. *(Jim Kinkaid)*

GATX 28382
DOT111A100W1
▲ A similar car photographed in June 1998. It was one of two cars (GATX 28381-28382) leased to Howell, who was based in Channelview, Texas. The car was constructed in June 1987.

(Emery Gulash, Morning Sun Books collection)

GATX 29159
DOT111A100W1
◄ Back to kaolin service cars. This example was at Kansas City, Missouri on April 17, 1994. It was part of the set 29151-29170, built in June 1977. The cars were formerly marked for the Georgia Kaolin Company.

(Jim Kinkaid)

GATX 40262
DOT105A500W
▼ Here's a photo taken on March 6, 1994 at Slidell, Louisiana. In lease service to the Olin Company, it was rated at 17,393 gallon capacity. The car belongs to the GATX series 40250-40349 which had been assembled in December 1969 through February 1970.

(Lee Yoder)

GATX 40416
DOT111A100W1
▲ Running around without any light weight information, it was at Birmingham, Alabama in August 1974. Assembled in July 1970 and fitted with heater coils, it belonged to the group 40395-40424, furnished under lot 8921.
(Jim Thorington)

GATX 40438
DOT111A100W1
▶ Another Wyandotte car found in June 1979. It was part of the series 40425-40472 and constructed in July (as this car was) and August 1970. The car is marked as being of 10,418 gallon capacity. *(William Rosenberg)*

GATX 49801
DOT111A60W1
▼ From December 1977 through February 1978 General American built cars 49751-49815, which included this car. Many were under lease to Cargill. The car illustrated here was assembled in January 1978 and was caught on film in April 1982 while at Memphis, Tennessee.
(Al Lanier, Jim Kinkaid collection)

GATX 51284
DOT111A100W4
▲ Likely built in May 1970, it was photographed in May 1972. The car was part of the set 51277-51291 and was of 20,889 gallon capacity. *(William Rosenberg)*

GATX 52547
DOT111A100W1
◀ Auburndale, Florida was the location of this car on May 11, 1984. The October 1981 date is when the 17,670 gallon car was manufactured at General American. It was part of the group 52539-52599, which included cars constructed in November and December 1981.
(Emery Gulash, Morning Sun Books collection)

GATX 58033
DOT111A100W
▲ Here we're at Altoona, Pennsylvania in June 1990. Constructed in September 1980, it was probably part of the set 58010-58048, though more cars may have been involved. Lou Ana, the company leasing the car (founded in 1894) was a large producer of vegetable oils.
(Jim Kinkaid)

GATX 65706

▼ This car is at Tehachapi, California on January 12, 1996 in sodium hydroxide and sodium peroxide service. It may have been built in December 1984 and the series is thought to be 65706-65711. *(Pat Holden, Jim Kinkaid collection)*

GATX 72076

▼ A kaolin car at Tullahoma, Tennessee in August 2003. At the time of the photo the coil equipped car was listed as part of the series 72070-72086. This series first appeared in the equipment register in April 1996 and the car may have been constructed the previous October. *(Lee Yoder)*

GATX 73490

AAR211A100W1

▼ Another car at Tullahoma, Tennessee. It was found there in May 2007 and was assembled by Trinity Industries in October 1998. The car, furnished under builder's file 9845, was part of the set 73241-73650. *(Lee Yoder)*

GATX 91597
ICC111A100W1
▲ A smaller car for corn syrup service at Birmingham, Alabama. It was there in April 1980, still with ICC specification markings. The coil equipped car, which belonged to the group 91573-91622, was assembled in February 1968.
(Jim Thorington)

GATX 91974
DOT111A100W1
◀ A car seen in Jessup, Maryland in December 1978. It had been assembled in September 1968 as part of the set 91949-91998, though only 46 cars were in the set. These cars had heater coils. *(Jim Rogers)*

GATX 92135
ICC111A100W1
▲ A car located at Birmingham, Alabama in March 1969. Assembled by General American in September 1968 it was under lease to a company incorporated in 1965 at Midland, Texas fur sulfur service. This particular car was placed within the 92124-92145 set in the equipment register. These cars had heater coils too. *(Jim Thorington)*

GATX 92978
ICC111A100W1
▶ Also at Birmingham, but in August 1970. With a 12,604 gallon capacity it was built by General American in August 1969. Although the car belonged to the set 92966-92995 in the equipment register, Kaiser may have only leased cars 92976-92991. Again, the car is fitted with heater coils.
(Jim Thorington)

GATX 93947
DOT111A100W1
▼ This car was put on film in May 1972 and used in glycerine service. Manufactured in December 1969 it was assigned to the series 93945-93949.
(Rail Data Service, Jim Kinkaid collection)

GATX 97009
DOT105A500W
▲ Used for metallic sodium service, this car was photographed in September 1972. It was constructed in December 1966 as one of ten such cars in the group 97005-97014. *(Paul C. Winters)*

GATX 300528
DOT111A60W1
◀ A final General American car in kaolin service. It was passing through Kansas City, Missouri on April 17, 1994. Built in February 1980 it belongs to the set 300527-300535 and was a 14,097 gallon car. *(Jim Kinkaid)*

GCX 410404
DOT111A100W2
▼ A photo taken in December 1978 of an Allied Chemical car assembled by Union Tank Car in November 1967. It was in sulfuric acid service and was part of the 10-car set 410400-410409, which were 10,000 gallon capacity cars. *(Paul C. Winters)*

HARX 1003
AAR211A100W1
▼ Minority carbuilder Hargis railcar, of Gordon, Georgia probably built less than 100 tank cars during its existence, with one being shown here. Found in September 1998 this car was constructed in March 1989 as part of the series 1000-1009. It is thought that the cars may have been assembled from Union Tank Car kits. *(Emery Gulash, Morning Sun Books collection)*

HCPX 1289
DOT105A500W
▶ Hauling chlorine, it was part of Hooker Chemical's series 1229-1289. As mentioned in Volume One, HCPX cars were leased via the Commonwealth Plan to Hooker. American Car and Foundry provided these cars under lot 18-13429 with the set being assembled in July 1968. Repainted from classic orange and black the car was at Kansas City, Missouri on April 16, 1994.
(Jim Kinkaid)

HCPX 4021
DOT103W
▼ Another Commonwealth Plan car, apparently assembled in February 1965. Used to carry phosphorus it was caught on film in May 1972. The cars, assigned the numbers 4001-4024, were fitted with heater coils and of 13,400 gallon capacity. The cleanout fittings at the tank ends are worthy of note. *(Paul C. Winters)*

HOKX 8407
DOT105A500W
▶ An interesting quandary on the Hooker Chemical chlorine cars is shown here and opposite page top. Both were assigned into the 111-car series 8358-8468, all of which first appeared in the April 1982 equipment register. Note that car 8407, a 17,547 gallon car, was manufactured in December 1981 by ACF. Its photo was taken at Magnolia, Arkansas on October 13, 1982.
(Al Lanier, Jim Kinkaid collection)

HOKX 8424
DOT105A500W
◀ The other part of the question deals with this car. Of 17,352 gallon size, it was provided by General American in January 1982. It is unknown how many of each builder were in the overall group. This picture was taken at McNeil, Arkansas on October 13, 1982.
(Al Lanier, Jim Kinkaid collection)

KPCX 705
DOT111A60W1
▼ Supplied to the Koppers Company by General American in January 1974 this car crawls through Lindsborg, Kansas in November 1988. It is obvious that the first try with stenciling the company logo failed due to spillage so it was moved. The 23,700 gallon car was part of the seven car set 700-706.
(Jim Kinkaid)

MAPX 12005
DOT112J340W
◀ This car was most likely assembled by North American Car at Texarkana, Texas. Assembled in the 1960's it was acquired used in late 1980 by Mapco Gas Products, the owner of the reporting marks. There were 15 of these numbered 12000-12014. It was found in Arkansas City, Kansas in December 1989. *(Jim Kinkaid)*

MGSX 151
DOT111A100W3

▶ A car operated by Martin Gas Sales at Mulberry, Florida on January 28, 1984. It was assembled by Trinity Industries at their Longview, Texas plant and is a 13,483 gallon capacity car. There were 65 of these 14,000 gallon cars and they were numbered 100-164.

(Emery Gulash, Morning Sun Books collection)

NATX 13604
DOT111A100W1

▼ Here is a car assembled by North American Car, with its tank supplied by the company AMF Bierd, in February 1971. It was in Birmingham, Alabama in August 1972 and was part of the 40-car set 13600-13639. They were fitted with heater coils. *(Jim Thorington)*

NATX 17210
DOT111A100W3

▼ While this car and 17213 (opposite page top) seem identical at first glance, they are subtly different. Close inspection of 17210 reveals it to be stenciled with one inch reporting marks and numbers above the weigh data. It was in Birmingham, Alabama in April 1980. *(Jim Thorington)*

NATX 17213
DOT111A100W3
▲ Here is one with proper stenciling. It was found at Memphis, Tennessee on March 20, 1983. Both cars were likely acquired used. The equipment register indicated that there were some 13 cars scattered within the number 17200-17231 series. As they were constructed by Union Tank car, they may have come from that source. *(Al Lanier, Jim Kinkaid collection)*

NATX 21862
ICC111A100W1
▲ This illustration was taken in April 1969 and the car was assembled at Texarkana, Texas by North American Car in August 1967. Fitted with heater coils it was one of 16 numbered as 21849-21864. *(Paul C. Winters)*

NATX 71126
DOT111A100W1
◀ Another car with heater coils, photographed on April 27, 1973 and apparently at the Kalama Chemical facility at Kalama, Washington. Assembled by North American Car at Texarkana in August 1971 it was part of a 250-car group numbered 71100-71349. The car had a capacity of 20,732 gallons. Kalama was a large producer of chemicals used in the manufacture of preservatives, flavorings and fragrances.
(G. Bolinsky, Jim Kinkaid collection)

NATX 190093
AAR211A100W1
▲ There were 120 of these cars assembled by Union Tank car in July 1998 under lot 5355. One was photographed on August 24, 2004 at Baltimore, Maryland. It was a part of the 190000-190119 series. The name "Cerestar" refers to a company owned by the Cargill Corporation that makes starch products. *(Jim Rogers)*

NCTX 33338
DOT112J340W
▶ NCTX was the reporting mark for North American's Canadian subsidiary. One of their cars is found here in July 1985. It was constructed by Hawker Siddeley in June 1972 at the Trenton, Nova Scotia plant. The car carries 28,263 Imperial gallons of anhydrous ammonia and was one of 341 cars in the set 33000-33341.
(Emery Gulash, Morning Sun Books collection)

OCCX 2121
DOT111A100W1
▼ Memphis, Tennessee was where this Occidental Chemical Company car was seen on February 12, 1986. Assembled in January 1981 as a 13,629 gallon car, it was from the 2001-2196 series.
(Al Lanier, Jim Kinkaid collection)

OLNX 22034
DOT111A100W3
▼ The Olin Corporation owned the car found in this illustration. It was at Birmingham, Alabama in October 1984. The car, part of the 86-car set 22001-22086, was built in December 1977. *(Jim Thorington)*

PQX 908
ICC111A100W1
◀ A photograph taken in June 1968 of a car manufactured in September 1967. Assembled by Union Tank Car it was one of 10 similar cars owned by the Philadelphia Quartz company, being numbered as 900-909.
(Rail Data Service, Jim Kinkaid collection)

PROX 13631
CTC111A100W2
▼ In June 1998 this car was caught on film, having been fabricated in December 1988. Of note is the dual French-English stenciling. Also of note is the "CTC" prefix to the specification, which stands for the Canadian Transport Commission, the Canadian equivalent to the ICC. Carrying sulfuric acid, it was likely from the set of cars numbered 13571-13680.
(Emery Gulash, Morning Sun Books collection)

PROX 28040
CTC105A500W
▲ Another Procor car at Kansas City, Missouri in chlorine service. It is displaying a January 1988 new date. The car is part of the 28001-28049 set. *(Jim Kinkaid)*

PROX 60140 & 60177
TC111A100W1
▲ ▶ Two cars in sulfur service built by Union Tank Car. At middle is one furnished in April 1993. It was at Reading, Pennsylvania on August 1, 1993. Below that is one with a different hue of yellow when photographed in October 1998. It was constructed in June 1993. Both cars were part of the PROX 60135-60184 series and were of nominal 13,200 gallon capacity. Note the Transport Canada specification markings.
(middle, Craig Bossler; bottom, Emery Gulash, Morning Sun Books collection)

PROX 70308
CTC111A100W1
▼ Another Procor car found in Tulsa, Oklahoma on January 11, 1992. The markings are hard to read but appear to indicate that the car was assembled in January 1985. At the time of the photo the car was part of the 70271-70404 group. *(Jim Kinkaid)*

PROX 75318
AAR211A100W1
▲ At Tullahoma, Tennessee in June 2001 is this "Opti-Miser" car. Note that it is marked as being under nitrogen pressure. *(Lee Yoder)*

PROX 83509
CTC105A500W
◀ In October 1997 the photographer found this car which had been built in May of that year. It is of 17,404 gallon capacity and carried chlorine. According to the equipment register, it was one of two such cars, the other being 83508.
(Emery Gulash, Morning Sun Books collection)

PROX 98632
CTC112J340W
▶ Our last Procor car is this monster photographed in July 1984. Assembled in November 1980 it was listed as part of the group 98624-98632 in the equipment register.
(Morning Sun Books collection)

RAIX 6462
DOT111A100W1
▼ U.S. Leasing International held these reporting marks when the car was found at Arkansas City, Kansas on May 11, 1991. Built in January 1976 and under lease to Union Carbide, it was one of 99 assigned the number series 6401-6499. The 20,000 gallon cars were manufactured by General American and deliveries ran from October 1975 through February 1976.
(Jim Kinkaid)

SHPX 206335
AAR211A100W1
▶ In August 2005 this tank car was at Tullahoma, Tennessee. A 16,251 gallon car it was part of the 125-set numbered 206276-206400, built about one year earlier. The commodity carried, limestone slurry, has one interesting use: it is used as a scrubber to clean coal fired power plant exhaust systems.
(Lee Yoder)

SHPX 207320
▲ In March 2005 this 24,198 gallon car was also at Tullahoma. It was listed as part of the 207318-207402 group and leased to Invista, a producer of polymers and polyesters. The commodity, dimethyl terephthalate, is a constituent of these products. *(Lee Yoder)*

SHPX 221231
DOT112J340W
▼ At South Belthlehem, Pennsylvania in July 2004. Assembled in July 2001 it was under lease to the Premcor Refining Group, who had refineries both in Texas and Tennessee. The car was one of 45 in the set 221216-221260. *(Eric Neubauer)*

STSX 1801
DOT111A100W1
◀ Trinity Industries' Longview, Texas plant produced this car under file 9020. The equipment register shows that it was part of a 583-car group numbered 1500-2082. The 17,600 gallon cars were delivered between August 1990 and February 1991. Here is a photo taken at Tullahoma, Tennessee in March 2006.
(Lee Yoder)

STSX 4042
AAR211A100W1
▲ Another Staley car found at Tullahoma, but on April 2, 1999, having been built by Trinity Industries the month before. It was a 19,564 gallon car fitted with heater coils. There were 750 of these numbered 4000-4749, supplied under builder's file 9889. *(Lee Yoder)*

TASX 7084
▶ In yellow while at Slidell, Louisiana on January 29, 1994 is this sulfur service car. It was part of the 193-car set listed under numbers 6920-7112. The coil-equipped cars were first reported in the April 1993 issue of the equipment register for owner Transammonia, Incorporated. *(Lee Yoder)*

TCPX 70009
▼ Another sulfur car found at Decatur, Alabama on April 6, 2002. Apparently assembled by Trinity Industries, it was part of a 100-car series first listed in April 2001 as TCPX 70000-70099. The cars were operated by Transport Capitol Rail Partners. *(Lee Yoder)*

TILX 135535
DOT111A100W1
▼ Continuing with sulfur service cars we see this one found in Tullahoma, Tennessee in February 2001. The car illustrated here was manufactured by Trinity Industries for their leasing operation and part of the 135520-135619 group. They were constructed in December 2000 under file T00063. *(Lee Yoder)*

TILX 170000
DOT111A100W3
▲ Down the road at Memphis was this car on March 20, 1983. One of 25 such cars built by Trinity, it was delivered under file 8048 as cars 170000-170024. Construction began in May 1981 and continued the following month or two. These were 17,600 gallon cars assembled at Tulsa, Oklahoma.
(Al Lanier, Jim Kinkaid collection)

TLDX 71708
DOT105A500W
◀ A car assembled by Richmond Tank Car in August 1969 for sale to Pullman-Standard's Transport Leasing Division. It part of a set then leased to the GAF Corporation as numbers 71700-71719. The car, seen in this photo dating to July 1972, was assembled under builder's lot 1735.
(William Rosenberg)

TLDX 71721
DOT105A500W
▲ Another car from Richmond Tank Car's lot 1735. Although insulated, the 17,300 gallon car lacked heater coils and was part of the series 71720-71724, which were furnished in December 1969. *(Pullman-Standard photo, Jim Kinkaid collection)*

TLDX 210000
DOT111A100W3
▶ Another builder's view. This example was one of four procured as 10,400 gallon sized cars in September 1971. Part of the series 210000-210003, they were insulated and had heater coils. The lessee was a company formed in 1954 as a joint venture between the Monsanto and Bayer companies and based in Pittsburgh, Pennsylvania to make polyurethanes.
(Pullman-Standard photo, Jim Kinkaid collection)

TLDX 216040
DOT111A100W1
▼ This view shows a 16,000 nominal capacity car with heater coils. It was part of the 21-car group 216036-216054 assembled in October 1968.
(Pullman-Standard photo, Jim Kinkaid collection)

TLDX 216081
DOT111A100W1
▲ Part of the 32-car series 216076-216107, this picture was taken during delivery in July 1969. These coil equipped cars had a capacity of 16,300 gallons. *(Pullman-Standard photo, Jim Kinkaid collection)*

TLDX 216091
DOT111A100W1
▲ Also from that group is this photo taken by the builder. The photo was shot in September 1969.

(Pullman-Standard photo, Jim Kinkaid collection)

TLDX 216112
DOT111A100W1
◀ Another production photograph which was taken in May 1970. The car in this illustration was one of 20 delivered as 216108-216127. The cars in this set were 16,300 gallon capacity cars and used coils.

(Pullman-Standard photo, Jim Kinkaid collection)

UELX 78016
DOT111A100W1
▲ At Decatur, Alabama in June 1978, this car is exactly one year old. It was assembled by American Car and Foundry at Milton, Pennsylvania in June 1977 under lot 18-15277. The owner, United Equipment Leasing Associates, operated 36 of these 21,000 gallon cars numbered 78003-78038. *(Bernie Wooller)*

UELX 78675
DOT111A100W3
▲ The Evans Company also built tank cars, as witnessed by this one photographed by them at their Junction City, Kansas plant in August 1980. There were 100 cars in this order, which was builder's lot 1765-01. They were numbered 78650-78749.

(Evans Company photo, Jim Kinkaid collection)

UTLX 13074
ICC111A100W1
▲ The car shown here (along with the 13096 at opposite top) may be somewhat unique. It was built in August 1957 with the tank constructed by Graver Tank and Manufacturing, but it would appear that Union Tank Car modified the cars in the early 1960's to a stub sill design. Car 13074, whose picture was taken in October 1969, was part of the group 13052-13074, which were 11,000 gallon cars. (The car to its left was ACFX 88930, profiled on page 5.) *(Paul C. Winters)*

UTLX 13096
ICC111A100W3
▶ Another car thought to have been modified by Union Tank Car, in this case probably in July 1963. It was assembled in October 1957 and was in muriatic acid service when caught on film. For this it was rubber lined. It was one of 59 cars placed within the 13092-13249 number series and was of 10,866 gallon capacity. *(Bernie Wooller)*

UTLX 23763
ICC111A100W3
▼ A car assembled in April 1966 and put onto film in May 1967. It would appear to have been part of a four car add-on (23762-23765) to the existing number series 23766-23774. The car has a 23,129 gallon capacity.
(Emery Gulash, Morning Sun Books collection)

UTLX 23771
ICC111A100W3
◀ Another car under lease to Dow in a view taken in May 1967. UTLX 23763, shown above, is coupled next to it. Assembled in July 1964 the car was of 23,168 gallon capacity and assigned into the 23766-23774 number series.
(Emery Gulash, Morning Sun Books collection)

UTLX 24769

DOT111A100W3

▲ The photographer found this car in September 1994. It was a 14,150 gallon car fabricated in April 1981 and as such it was part of the set 24739-24769. KT Clays was more properly known as the Kentucky-Tennessee Clay company, a kaolin producer.

(Emery Gulash, Morning Sun Books collection)

UTLX 24887 & 24949

DOT111A100W1 & W3

▶ ▼ Two cars from the equipment register series 24787-24993. A middle is a kaolin clay slurry car in Kansas City, Missouri on April 17, 1994. The car had been furnished in September 1980 as a 14,115 gallon car. Below that is is a 13,801 gallon limestone slurry car found at Laurel Siding in Pennsylvania on October 3, 1982. It was fabricated in December 1979.

(middle, Jim Kinkaid; bottom, Craig Bossler)

UTLX 25983

DOT111A100W1

▼ The car shown here was part of the four car set 25983-25986, built in May 1979. Its photo was taken in January 1987. The car had a capacity of 14,125 gallons. *(Emery Gulash, Morning Sun Books collection)*

UTLX 42028 & 42086

AAR206W

▲ ◀ At the middle position is a car at Birmingham, Alabama in March 1969. It is carrying superphosphoric acid (a product used as a fireproofing agent) while under lease to the Tuloma Gas Products Company and was fabricated in January 1967. Below is a similar car in Birmingham found in April 1970. It was assembled in March 1967 and records indicate that it had a capacity of 12,478 gallons. Both cars were part of the UTLX series 42000-42116. Note the oversized dome and outlet housings on these cars which allowed extra-heavy insulation. They are another of those rare "tank-within-a-tank" designs (see DUPX 20034 on page 51).

(both, Jim Thorington)

UTLX 43331
DOT111A100W1
▲ A photo taken in August 1987 at Baltimore, Maryland of a 13,655 gallon car. In the equipment register it was denoted as part of the 43310-43345 group. *(Jim Rogers)*

UTLX 65839
ICC111A100W3
▶ Fitted with heater coils for sulfur service, the car was assembled in February 1963 with a tank constructed by Graver Tank and Manufacturing. Its photo was taken in January 1987. According to the equipment register the car was assigned into the 65775-65841 number set.
(Emery Gulash, Morning Sun Books collection)

UTLX 66432
DOT111A100W1
▼ Here's a photo taken in June 1983 at Paducah, Kentucky. The car was of a 16,440 gallon capacity.
(Al Lanier, Jim Kinkaid collection)

UTLX 71899
DOT111A100W
▲ Manufactured in May 1977 was this 17,577 gallon car. It was found at Colton, California in June 1985. At the time of the picture it was recorded in the equipment register as being part of the 71875-71908 group of cars. *(Pat Holden, Jim Kinkaid collection)*

UTLX 72257
DOT111A100W3
◀ Rusty tracks, rusty wheels: not much has been moving at this location in Reading, Pennsylvania for some time. The car was photographed there on May 31, 1982. Fitted with heater coils and assembled in December 1975 it was of an 8,176 gallon capacity. Information suggests that it was part of the series 72202-72291. *(Craig Bossler)*

UTLX 74206
DOT111A100W3
▼ More rusted wheels at Bakersfield, California on September 22, 1985. The car was described as having a capacity of 23,544 gallons and probably part of the series 74166-74251. *(Pat Holden, Jim Kinkaid collection)*

UTLX 74841
DOT111A100W1
▲ A car on the move at Kansas City, Missouri on March 26, 1992. Note that the car was in citric acid service and built in February 1970. It was of a 17,140 gallon capacity. *(Jim Kinkaid)*

UTLX 75031
DOT111A100W1
▲ Union Tank Car called this a "Changeable Car" since it was designed to be able to have its size changed by adding or removing tank barrel sections. This one was manufactured in May 1970 and was at Newton, Kansas on September 9, 1990. Information shows that the car, of 23,484 gallon capacity, was likely to be part of the series 75000-75099, all of similar size. *(Jim Kinkaid)*

UTLX 76246
DOT111A100W
▶ The photographer caught this 20,488 gallon car on July 4, 1973. The car has heater coils and was listed as part of the five car set numbered 76245-76249.
(William Rosenberg)

UTLX 77310
DOT111A100W3
▲ The Arkansas Grain Corporation, a maker of soy-based salad oils, was using this car when its photo was taken in July 1970. Constructed in May 1969 it carried 20,472 gallons of commodity. By October 1969 it was described as being part of the 77299-77312 set of cars.
(Paul C. Winters)

UTLX 77322
DOT111A100W3
▲ An illustration taken in June 1972. Like UTLX 77310 above, it was assembled in May 1969, and by that October was listed as part of the 77322-77330 series. It carries 20,462 gallons of product (the lessee was a producer of oils and shortenings). *(Jim Thorington)*

UTLX 78810
◀ This picture was taken in April 1971. Close examination reveals the insulation jacket damaged on the far end of the car. According to records, the car carried 16,400 gallons (it was likely in chlorine service) and was probably part of the 78809-78819 number set.
(William Rosenberg)

UTLX 78868
ICC111A100W3
▲ This illustration was taken in December 1978. The Changeable Car was assembled in May 1968 and from the looks of things, apparently in asphalt service. The car carried 22,741 gallons and was described as being part of the number series 78820-78894. *(Paul C. Winters)*

UTLX 79829
ICC111A100W3
▼ Another unusually arranged car found in March 1980. It was assembled in April 1967 and was in corn syrup service. The car, with a rating of 8,326 gallons, was part of the 79825-79899 group of cars. Additional cars were fabricated in March and June with at least one additional company (Rohm and Haas) leasing cars from this set. *(Paul C. Winters)*

UTLX 82117
DOT105A400W
▲ A nice down-on shot taken at Kansas City, Missouri on the Burlington Northern tracks in September 1990. Note the converted friction bearing trucks, soon to be outlawed in interchange service. Under lease to ICI Americas, Incorporated, it was of 11,069 gallon size and probably part of the 82105-82117 set of cars. *(Jim Kinkaid)*

UTLX 85021
DOT111A100W3
◀ A car at Ft. Meade Junction in Maryland is shown in a picture taken in July 1988. Although rather plain, its mud shields above the wheels are noteworthy, as is the offset manway and friction bearing trucks. The car belongs to the set 85010-85029 and was under lease to the National Starch company. *(Jim Rogers)*

UTLX 85377 & 85411
DOT111A100W3
▲▼ Two cars with unusual highly offset manway locations. Both cars were from the series 85375-85485. At middle is a car in prune concentrate service (of all things!) found sitting at Reading, Pennsylvania on April 14, 1986. Coil-equipped, it was manufactured in April 1965 and carried 8,320 gallons of concentrate. And below is a car seen at Birmingham, Alabama in April 1980. It was constructed in August 1965 and carried 8,307 gallons. Given the company leasing the car, it was likely carrying corn syrup. *(middle, Craig Bossler; bottom, Jim Thorington)*

UTLX 300163
DOT111A100W3
▶ A photo taken on October 4, 1992 at Schiller Park, Illinois. Manufactured in October 1986 it was one of 50 in the set 300145-300194. The cars were assembled in October and November of that year under lot 6419. *(Gib Albach)*

UTLX 300519
DOT111A100W1
▼ There were 22 of these titanium dioxide service cars fabricated in June and July 1987 and numbered 300500-300521. This one was photographed on May 22, 1992 at Richmond, Virginia. It was an 11,316 gallon car.
(Gib Albach)

UTLX 643548
▲ A car in toluene diisocyanate service (a constituent of polyurethane manufacturing) is found at Decatur, Alabama on April 6, 2002. It was one of 52 cars numbered 643546-643597, all formerly in Olin lease service and constructed in December 1992 under lot 6906.
(Lee Yoder)

UTLX 650912
DOT111A100W1
▲ Here is a rather unusual car. It would appear to have electric heaters yet seems to be also fitted with heater coils. The car, with a 19,024 gallon capacity, was part of the 60-car set 650873-650932. Under lease to Koppers Incorporated it was spotted at Tullahoma, Tennessee in May 1998. *(Lee Yoder)*

UTLX 651107 & 651108
▲ ▼ Two cars from the 64-car group 651103-651166, which first appeared in the July 1995 issue of the *Official Railway Equipment Register*. Apparently acquired second hand and recently repainted, observe the difference in stencil colors. Both were spotted in Tullahoma, Tennessee, the middle one in April 2006 and the lower one in June 2004. *(both, Lee Yoder)*

UTLX 802735
DOT111A100W3

▶ A car originally owned by Richmond Leasing and acquired used by Union Tank Car. It was constructed by Richmond Tank Car in June 1978 as RTMX 2735, which was part of the series 2727-2741 (Union Tank Car simply added the prefix "80" to the car numbers when they acquired them). The car carried 20,859 gallons and was seen at Kansas City, Missouri in 1995. *(Jim Kinkaid)*

UTLX 900828
DOT105A300W

▼ Nicely painted while in Tulsa, Oklahoma on January 11, 1992 while in propylene oxide service (an organic compound utilized in the manufacture of polyurethane plastics). Observe how the ladder is inset into the insulation jacket to maintain Plate "C" clearances. It was one of 106 numbered 900747-900852 and constructed in June 1991. It is believed that all cars were built under lot 6805 for lease to Arco. *(Jim Kinkaid)*

Multi Compartment Cars

Not every car presented in this section is truly a multi-compartment car. Generally speaking the number of safety valves directly relates to the number of compartments in a tank car. When known, I've provided this information. But in all cases, the cars are here to prevent confusion with "typical" single compartment cars shown elsewhere.

ACFX 71593
DOT111A100W6

▲ As it happens, our first car in this section is not really a multi-compartment car. Yet the multiple manway covers and hoods certainly seem to indicate otherwise. Leased to Rhom and Haas and equipped with coils, it was found in Reading, Pennsylvania on September 11, 1989. The insulated car belongs to the set 71591-71599 and was assembled in December 1986. *(Craig Bossler)*

ACFX 73371
DOT111A100W1
▲ Another imposter. Fabricated in November 1989 for phosphorus service with the Monsanto Company, it is a single compartment insulated car with twin sets of loading hatches. Assigned into the 73357-73384 number group it was photographed in October 1998.
(Emery Gulash, Morning Sun Books collection)

ACFX 76386
DOT111A100W1
▲ Another car in phosphorus service. The small diamond symbol shows that it has one approved manway, thus it is a single compartment car, yet the profusion of covers and weather hoods makes it seem like a multi-compartment car. It was in Reading, Pennsylvania on May 5, 1987, having been completed by ACF in October 1984 and under lease to the FMC Corporation. The equipment register has it as part of the 76380-76406 group. *(Craig Bossler)*

ACFX 79818
DOT111A100W1
◀ A similar car. Also under lease to FMC it was in Reading, Pennsylvania on September 22, 1984. This was one of 18 cars assigned into the 79808-79818 number set. These cars were assembled in January and February 1981 and also January 1982. It was a 13,903 gallon capacity insulated car. *(Craig Bossler)*

ACFX 82777
DOT111A100W1
▲ Our first true multi-compartment car. This can be determined both by the dual outlets and also by the dual safety vents, located near the ends of the manway platforms. This non-insulated car was at Altoona, Pennsylvania in June 1990. It was leased to the Na-Churs Company (of Marion, Ohio) for use in carrying liquid plant food. It was supplied as part of the 20-car set 82775-82794, assembled in April and May 1969. *(Jim Kinkaid)*

ACFX 88869
DOT111A100W6
▶ Under lease to the Nabisco Company, the non-insulated car sits at Reading, Pennsylvania on August 9, 1984. Its three compartments carry 8,501, 8,406 and 8,497 gallons respectfully. It was manufactured in January 1969 with a stainless steel tank (alloy steel is the reason for the -6 suffix to the specification) and was part of a three car set listed as 88868-88870. *(Craig Bossler)*

ACFX 94650
DOT111A100W1
▼ Another insulated phosphorus service car caught on film in October 1998. It was one of 42 in the series 94646-94687 and was constructed in August 1992. Note the drip gutter and piping down the car centerline and the Monsanto logo.
(Emery Gulash, Morning Sun Books collection)

AMOX 911

AAR211A100W1

▼ A training car at Jessup, Maryland in March 1987. It was assembled by Union Tank Car in October 1972 as a two compartment uninsulated car fitted with heater coils and modified for Amoco in April 1987. Certainly unusual with its profusion of manway and outlets of various designs it was the only one of its kind. Note the starred capacity of zero pounds! *(Jim Rogers)*

CELX 2059

DOT111A100W1

▼ A two compartment car drifting downgrade at Altoona, Pennsylvania in June 1990. It carries 10,438 gallons in the left compartment and 10,450 in the right. The car was one of five in the group 2055-2059, assembled by Richmond Tank Car in February 1970. The insulated cars had heater coils. *(Jim Kinkaid)*

DOWX 3180

AAR211A100W1

▲ This unusual car was sitting in Wichita, Kansas in March 1990. It is a one compartment car, as denoted by the single outlet. The eight manway hatches are certainly unusual and the 28,600 gallon non-insulated car has a sparger system installed. Fabricated by General American in March 1979 it was one of 32 in the set 3164-3195. *(Jim Kinkaid)*

DUPX 14011
AAR211A100W1
◀ Another sparger equipped car. It is also a single compartment car, but fitted with four manway hatches. Found in Memphis, Tennessee on August 27, 1982, there were 17 of these cars numbered as 14001-14017, built at General American in May 1967.
(Al Lanier, Jim Kinkaid collection)

GATX 16485
DOT111A60W1
▶ A single car order resulted in this one being assembled in April 1974 as a two compartment insulated car. The far (or "A") tank holds 4,150 gallons while the near one (obviously called the "B" tank) carries 4,145. The coil equipped car was at Muhlenberg Township, Pennsylvania on March 5, 1977. The company leasing the car was incorporated in 1923 and was an industrial cleaner manufacturer.
(Craig Bossler)

GATX 22245
DOT111A60W1
▲ Here's a two compartment car at Jessup, Maryland in December 1981. Although the uninsulated, coil equipped car is plainly painted, note the friction bearing trucks and all the fittings crowded together on the top side of it. The equipment register has the car listed as part of the 22244-22250 number group. It was assembled in March 1967. *(Jim Rogers)*

GATX 31650
DOT111A100W1
▲ Here is one half of a two car set (the other car being GATX 31649) constructed in November 1982. On September 29, 1991 the five compartment car was at Kansas City, Missouri just exiting the Burlington Northern yard there. Presumably it is used to move different grades of lubricating oil. It has coils and is non-insulated. *(Jim Kinkaid)*

GATX 41108
DOT111A100W1
▼ Under lease to the Occidental Chemical Corporation when spotted at Reading, Pennsylvania on March 26, 1991, this was a single compartment car carrying 14,640 gallons. Having three manway hatches the uninsulated car was constructed in December 1970. At the time of the photo it was listed in the register as one of nine numbered 41101-41109. *(Craig Bossler)*

HOKX 70022
DOT111A100W1
◀ Found in Tullahoma, Tennessee in July 2001, this was a two compartment non-insulated car with eight manways. Several items are of note here. The car has continuous drip gutters completely surrounding the hatches that feed to a discharge pipe running to ground next to the ladder. Also of interest is the walkway support structure that remains level even though the tank body slopes to center, making the supports of varying height. It was assembled in February 2000 and from the 70015-70031 set. A most unusual car! *(Lee Yoder)*

NATX 24350
DOT111A100W1
▶ Another two compartment non-insulated car that was photographed in March 1991. The car was constructed by North American Car at Texarkana, Texas in August 1971. The far (or "B") compartment has markings showing its capacity is 10,335 gallons while the near one is rated at 10,347 gallons. It was part of the 24350-24365 car set.
(Jim Kinkaid collection)

NTLX 2302 & 2400
DOT111A100W1
▲▼ Two cars from minority car builder Texana Tank Car and Manufacturing of Nash, Texas. Both uninsulated cars were provided for their subsidiary Nashtex Leasing. At middle is a photo taken on October 14, 1998 at Reading, Pennsylvania. Here we see a car that has evenly divided tanks, each of 8,063 gallon capacity. It was constructed in January 1997 and one of two (the other car being 2301). Below that is another car also at Reading, but on May 7, 1999. This four compartment car was also constructed that January 1997. It was one of three numbered 2303, 2400 and 2402. Noteworthy on both of these cars are the bands welded to the outside of each individual tank where the tank heads join.
(both, Craig Bossler)

SHPX 203854
DOT111S100W1

◀ At Tullahoma, Tennessee in July 2006 was this phosphorus service car. Like the other cars used in this service and shown in this section, it is actually a single compartment car. The "S" within the specification is quite unusual for a general service car, signifying that it has head shields. The 13,231 insulated gallon car was part of a 52-car set numbered 203821-203872 and under lease to P4 Productions, which was a joint venture between Monsanto and another firm based out of Soda Springs, Idaho. These cars were assembled in December 2000 and January of the following year. *(Lee Yoder)*

UTLX 1032 & 1036
DOT/ICC111A100W1

▲ ▼Two cars from the UTLX group 1032-1040, added in late 1960 to the previously existing 1041-1049 set. The photos show the cars both as delivered and as updated. At middle is the updated car while at Kansas City, Missouri in August 1989. This two compartment non-insulated car had capacities of 10,220 gallons on the left compartment and 10,231 on the right. Note that the side walkways and railings have been removed, leaving just the mount brackets. Below is a car seen August 1963 illustrating its original configuration. It was constructed in May 1960.
(middle, Eric Neubauer; bottom, Paul C. Winters)

UTLX 1069
ICC111A100W1

▼ Here is a car fabricated in June 1962 by Union Tank Car. Unlike the two illustrated previously it originally came with end-mounted running boards. It was caught on film at Columbus, Ohio on December 9, 1962. It's "A" compartment holds 10,289 gallons while the "B" one holds 10,283. The car was one of four listed in mid-1963 as numbers 1069-1072 which were then added to the previously existing 1051-1068 set. *(Paul C. Winters)*

UTLX 4499
ICC111A60W1

▶ This is a picture of a three compartment car built in May 1967, found in Reading, Pennsylvania on September 2, 1993. The compartments are numbered and from far end to near the capacities are: (1) 10,321 gallon; (2) 5,152 gallon and (3) 5,122 gallon. This coil equipped, non-insulated car was a single car order. *(Craig Bossler)*

UTLX 4525
ICC111A100W3

▼ Another three compartment car. Insulated and coil equipped it was assembled in November 1963. Its compartments had capacities of: (A) 3,437; (B) 3,438 and (C) 3444 gallons each. The car was also at Reading, but on April 3, 1987. It was under lease to the Exxon Company and it too was a single car listing in the equipment register. *(Craig Bossler)*

UTLX 67635
DOT111A100W1
◀ Our last "multi-compartment" car is like the first: it is a phosphorus service car that is really a single compartment car. And like the others shown in this section, it has a profusion of fittings topside. Tullahoma, Tennessee was where this shot was taken in June 2005. The car carried 14,217 gallons and was insulated. *(Lee Yoder)*

The "Beer Can" Cars

There is no "real" definition of what constitutes a beer can car that I am aware of, but here I'll use the term to describe stub sill cars with very short wheelbases.

DOWX 4928
DOT111A100W5
◀ Our first beer can car is a 10,317 gallon car located in Muhlenberg Township, Pennsylvania on October 26, 1988. Assembled in April 1968 by General American it was rubber lined for zinc chloride service. Fitted with mud shields, the uninsulated car became part of the 4918-4929 number group. The "5" suffix to the specification means that the car lacks bottom outlets or safety valves, having safety vents instead. *(Craig Bossler)*

ECUX 102031
DOT111A100W1
▼ An almost identical car found at Houston, Texas on May 6, 1992. The reporting mark ECUX belonged to Exxon Chemicals Americas and the 10,000 gallon car was originally part of the Humble Pipeline (HPLX) fleet numbered 102001-102090, constructed in September and October 1967 by General American. *(Jim Kinkaid)*

GATX 9530
DOT111A60W1
▲ A nice car found in June 1968. Note the pipes welded to the tank head, apparently they are heater coils. The non-insulated car was assembled in August 1965 and part of the 9525-9534 set.
(Paul C. Winters)

GATX 10002
DOT111A100W1
▶ This insulated car was caught at Tullahoma, Tennessee in August 1968. It had been supplied in March 1968 and was apparently a single car order.
(Lee Yoder)

GATX 21686
DOT111A60W1
▼ A typical car found in Reading, Pennsylvania on January 4, 1986. It too has end pipes and the car was assembled in June 1965. At the time of its photograph, the equipment register had it listed as part of the 21679-21689 number series. *(Craig Bossler)*

GATX 55183
DOT111A100W1
▲ We are back in Tullahoma, where the photographer found this car on June 2, 1974, which was being leased by a large supplier of water treatment chemicals. An insulated car it was built in August 1970 and put into the GATX 55175-55199 set. *(Lee Yoder)*

GATX 88384
DOT111A60W1
◀ At Birmingham, Alabama in March 1972. The uninsulated car shown here was thought to have been constructed in September 1965. In the equipment register it was simply shown as somewhere within the 88355-88499 number series. Geigy, the lessee, was a large dye and pigment manufacturer. *(Jim Thorington)*

GATX 91380
DOT111A60W1
▲ Another Birmingham photo, but in March 1971. The car, which is insulated and has heater coils, was built in September 1968 and assigned into the 91378-91394 group. *(Jim Thorington)*

GATX 93493
DOT111A100W1
▶ Here's a car found at Temple, Pennsylvania on May 20, 1972. It is a 10,350 gallon car constructed in May 1969 and is non-insulated. The car was part of the 93487-93498 set.
(Craig Bossler)

GATX 93598
DOT111A100W1
▼ An insulated and coil equipped car found while at Reading, Pennsylvania on June 13, 1987. It is of 10,345 gallon capacity and one of ten in the 93590-93599 group. *(Craig Bossler)*

HOCX 147
ICC111A60W1
▲ Surprisingly enough, "HOCX" is the reporting mark of the Head On Collision Line. One example was in Council Bluffs, Iowa on August 17, 1969. Markings indicate that it was fabricated in October 1966. The 10,000 gallon car was part of the set 141-159, built by General American. *(Lou Schmitz)*

HOCX 155
DOT111A60W1
▲ Another located at Pine Bluff, Arkansas on July 19, 1982. This one was assembled in November 1966.
(Al Lanier, Jim Kinkaid collection)

JMHX 6027
DOT111A100W1
◀ We are at Birmingham, Alabama in March 1969. The J.M. Huber Corporation operated these cars which were insulated and had heater coils. The 10,000 gallon car illustrated here was built in November 1966 by General American and part of the 6000-6031 series.
(Jim Thorington)

NATX 8256
DOT111A100W1
▶ An 8,750 gallon car in Memphis, Tennessee on October 18, 1980. It was built in July 1972 as part of the 8250-8274 car set and is insulated with heater coils.
(Al Lanier, Jim Kinkaid collection)

NATX 10841 & 10857
DOT111A100W1
▲▼ Two cars forming a part of the NATX 10841-10865 number group. The upper one was at Tullahoma, Tennessee on November 13, 1983. It was supplied in September 1974. Meanwhile, car 10857 was at Baldwin, Florida in December 1979, apparently in original paint. It had been assembled by North American at the Texarkana plant in October 1974. These cars had heater coils and were insulated.
(top, Lee Yoder; middle, Emery Gulash, Morning Sun Books collection)

RAIX 5049
DOT111A100W1
▶ U.S. Leasing International originally listed the car as part of the RALX 5000-5054 set in July 1970. By October the reporting marks had been changed to RAIX. These 10,300 gallon cars were built in May 1970 by Union Tank Car. This one was at Hearne, Texas on June 23, 1988.
(Al Lanier, Jim Kinkaid collection)

TCX 1825
DOT111A60W1
◀ A Texaco Incorporated car sits in Birmingham, Alabama in July 1968, four months after its manufacture by General American. The 10,348 gallon car, which had coils and was insulated, was one of 30 in the series 1800-1829. *(Jim Thorington)*

UTLX 76917
DOT111A60W1
▼ We are in Memphis for a photo taken in September 1983. The car illustrated was assembled in July 1969 and part of the 76913-76919 set. It was of 10,239 gallon capacity. Note that even though a short car, it has the funnel flow design. NL Industries (formerly the National Lead Company, a company that can be traced back to 1891 and renamed in 1971) was a large manufacturer of titanium dioxide pigment.
(Al Lanier, Jim Kinkaid collection)

UTLX 78363
DOT111A60W1
◀ Another funnel flow beer can car. This one was in Tullahoma, Tennessee on October 6, 1974. The insulated car was of a 10,251 gallon capacity and had been built in July 1968. It was listed in the equipment register as part of the three car set 78362-78364.
(Lee Yoder)

UTLX 78364
DOT111A60W1
◀ Reading the caption for car 78363 on the previous page would lead one to suspect that with only three cars, all would be painted for American Maize. However, look at this 10,260 gallon car, found in Huntsville, Alabama in November 1968, just after its July manufacture date.
(Bernie Wooller)

UTLX 78504
DOT111A60W1
▶ From the series 78500-78589, the car had been furnished in August 1967. Insulated, it was seen at Detroit, Michigan. The car has a 10,237 gallon capacity.
(Emery Gulash, Morning Sun Books collection)

UTLX 85926
DOT111A100W1
▼ Another car found at Detroit. This one was a 10,219 gallon car assembled in February 1974. Insulated, it was one of 11 cars placed within the 85914-85927 series in the equipment register.
(Jim Thorington)

VMCX 1203
DOT111A60W
▶ This non-insulated car was part of the 1201-1204 number set, to which was added 1205-1207. These 10,000 gallon cars were built by General American in early 1968. As mentioned before, VMCX cars were leased to the Vulcan Materials Company via the Commonwealth Plan.
(Rail Data Service, Jim Kinkaid collection)

Whalebellies

Whalebelly cars, or dual-diameter cars, were popular in the 1960's as builders strived for more capacity within existing clearances. However, by the early 1970's this type of car design fell out of favor as larger straight barreled cars were introduced.

ACFX 16895
DOT112A400W
◀ Not every whalebelly car is a huge monster. Witness this shorty found in Little Rock, Arkansas on March 15, 1986. Carrying 18,250 gallons of hydrofluoric acid it was assembled in July 1966 as one of ten in the group 16895-16904 under lot 71-13215.
(Al Lanier, Jim Kinkaid collection)

ACFX 16950
DOT112S340W
▼ Anhydrous ammonia was being carried in the car when its photo was taken in November 1987. Lessee Royster Company was a major mid-western distributer and manufacturer of ammonia-based fertilizers. The car was part of a 57-car series numbered 16905-16961, first reported in the equipment register in April 1967. It had been constructed the previous month. *(Emery Gulash, Morning Sun Books collection)*

ACFX 17074
ICC112A340W
▲ From the set 17000-17097, it was spotted on the Southern Railway in Huntsville, Alabama in January 1974. This car was constructed in September 1964. *(Bernie Wooller)*

ACFX 18739
ICC112A340W
▶ Another short whalebelly car found in December 1978. It was constructed in July 1968 and under lease to the B.F. Goodrich Company for vinyl chloride service. As such it was part of a 23-car series numbered 18738-18747.
(Paul C. Winters)

BRKX 235
DOT112J340W
▼ Burke Energy, of Hutchinson, Kansas, first reported this car in the April 1982 edition of the equipment register, which was when the company also first began listing cars. It was one of two, the other being numbered 232.
(Dave Casdorph, Jim Kinkaid collection)

CNTX 1004
ICC112A300W
▲ The Continental Tank Car Corporation used these reporting marks in July 1962 when the car was first listed in the equipment register. It was part of the six car set numbered 1001-1006. The 30,300 gallon car, assembled by General American in July 1962, was put on film in September of that year. *(Rail Data Service, Jim Kinkaid collection)*

CSOX 31020 & 31034
ICC112A340W
▲ ▲ Two Cities Service (later Citgo) cars built in October 1963 by General American. The car at middle was spotted in March 1973. The one below was found at Tullahoma, Tennessee in November 1967 with an earlier Cities Service logo. Both came form the 31000-31049 number group and were in liquefied petroleum gas service. *(middle, Calvin T. Banse; bottom, Lee Yoder)*

GATX 84444
ICC112A400W
▶ While the slide mount was marked "9444" and the car seems stenciled "8444" it is almost surely actually 84444. Found in July 1963 it is likely from the GATX series 84200-84499, although the quantity of cars has not been resolved. *(Emery Gulash, Morning Sun Books collection)*

GATX 30752
ICC112A340W
▼ From a 35-car set is this one in vinyl chloride service when spotted in May 1966. This car was constructed in December 1964 and belongs to the series 30750-30784. *(Paul C. Winters)*

GATX 83299
ICC112A340W
▼ Thought to have been assembled in January 1966 and in anhydrous ammonia service, it was spotted in April 1969. *(Paul C. Winters)*

GATX 83654 & 83719
ICC112A400W
◄▼ Two cars used in liquefied petroleum gas service. The upper car was in Tullahoma, Tennessee in August 1963. It was likely fabricated in November 1961. At middle is a car assembled in April 1962 and found in Columbus, Ohio on September 5th of that year. Both were listed as part of a set of 33 cars assigned somewhere into the 83600-83759 number series.
(top, Lee Yoder; middle, Paul C. Winters)

MCVX 83639
DOT112S340W
▲ Here is another Massachusetts call/volunteer fire department training car. It was located at Baltimore, Maryland in April 2002. The car may be ex-UTLX 83639, assembled in November 1963 and probably converted in August 1979. Like the earlier MCVX car in this Volume, it was never listed in the equipment register. *(Jim Rogers)*

NATX 32117

▲ North American probably used this car in general lease service when it was found sitting in Tullahoma, Tennessee in September 1966. It was constructed by them in December 1964 as part of the 152-car set numbered NATX 32000-32151.

(Emery Gulash, Morning Sun Books collection)

NATX 33890

▶ It ended up belonging to a set of 939 cars, eventually to be numbered 33800-34738. However, the car, which was likely constructed in January 1966, appears to be part of a 51-car set delivered early that year numbered 33851-33901. In any event, it was photographed at Birmingham, Alabama in March 1969. Its lessee used the car to haul anhydrous ammonia in support of its agricultural fertilizer business.

(Jim Thorington)

PSPX 30030

ICC112A340W

▼ In August 1961 General American supplied Phillips Petroleum with 50 of these cars. Rated at 30,300 gallons capacity, they were numbered as 30001-30050.

(Dave McKay, Morning Sun Books collection)

PYFX 1024
ICC112A340W
◀ The photo of this car, taken in April 1972, is of a car from the Pyrofax Gas Corporation's 1021-1040 series. They were nominally rated 33,300 gallon capacity cars. The car was assembled by American Car and Foundry in October 1964.
(William Rosenberg)

SHPX 16993
▼ It is hard to tell where this photograph was taken, but it was marked "Providence." Rhode Island perhaps? But it is known that the car was put onto film on August 1, 1967. The car was built in April 1967 and listed as part of the 10-car set numbered 16989-16998.
(George Turnbull, Jim Kinkaid collection)

SHPX 17005
ICC112A340W
▲ There were 40 of these cars assembled as numbers 17000-17039. This is a photo of a car constructed in April 1964 and taken in September 1967. *(Paul C. Winters)*

TGPX 32801
ICC112A340W
▼ Huntsville, Alabama was where this car was found in March 1969. It had been assembled in June 1963 by General American and was a 32,802 gallon car. Tamak Gas Products, the owner, had 11 such cars numbered 32801-32811. *(Bernie Wooller)*

UPCX 83642
ICC112A400W
▶ Here is a 30,277 gallon car photographed in May 1962. It had been constructed just the previous month by General American. Union Petroleum owned 20 of these liquefied petroleum gas cars, numbered 83630-83649.
(Calvin T. Banse)

UTLX 80706
ICC112A340W
▼ What's so interesting about this car in its August 1971 photo is the fact that the tapered sections of the tank shell end up right at the tank heads. A look at most of the other whale belly cars here will reveal that there is a cylindrical section inserted between the tapered portion and the tank heads where the bolster resides. The car carried 33,804 gallons and at the time of its photo, was described as one of 476 such cars within the UTLX group 80500-80999. As may be seen it was assembled in December 1965. *(Paul C. Winters)*

UTLX 83767
ICC112A340W
▲ Another shot of one of these unusual cars taken in Birmingham, Alabama in May 1964. Built in October 1963 it was of 30,716 gallon size and was one of 246 listed within the 83700-83999 number set in early 1964. *(Jim Thorington)*

UTLX 99206
ICC112A340W
◀ Trenton, New Jersey was where this car was spotted in June 1968. It has a capacity of 33,665 gallons and was assembled in November 1964. In early 1965 the car was shown as belonging to the 99198-99262 group. *(William Rosenberg)*

WRNX 30057
ICC112A340W
▼ A Warren car found in August 1967. Built by General American in May 1964 as a 32,860 gallon car, it was part of 200-car series 30053-30252.
(William Rosenberg)

The Big Cars With 6 And 8-Axles

ACFX 88366
ICC111A100W1
▶ A six axle car found at Enola, Pennsylvania on February 29, 1976. Built in January 1968 it would appear to be part of the series 88362-88366. *(Craig Bossler)*

CCBX 4815
DOT114J340W
▼ Also at Enola, but on May 17, 1986, was this 48,330 gallon Union Carbide car. It was first listed in the equipment register in July 1984 as one of seven miscellaneous cars and was probably acquired used. For the curious, CCBX once was assigned to the Carbide and Carbon Chemicals Corporation, later folded into Union Carbide.
(Craig Bossler)

DUPX 28068
DOT105A500W
▶ The DuPont car illustrated here appears to have been assembled in March 1969 by General American. The car was part of the set 28061-28070 added at that time to the previously existing 28050-28060 set. It was photographed in April 1970. *(Paul C. Winters)*

DUPX 28081
DOT105A300W
▲ Another one in May 1975. The car shown here was built in March 1970 and formed part of the 28071-28090 group, which were manufactured by General American too. *(Paul C. Winters)*

DUPX 29502
ICC111A100W1
▲ This six axle car was put onto film in July 1968 at Birmingham, Alabama. There were 75 of these cars built by General American. The car, assembled in October 1967, was assigned to the 29500-29574 series. *(Jim Thorington)*

DUPX 29636
ICC111A100W1
▼ And another DuPont car at Birmingham, in April 1970. Note that its trucks differ from the one above. Assembled in August 1968 by General American, it was part of a 43-car set numbered 29600-29642. *(Jim Thorington)*

DUPX 29705 & 29760
ICC111A100W
▶ ▼ Two cars from the 70-car set DUPX 29700-29769. American Car and Foundry supplied the cars in late 1968 and early 1969 for ethylene glycol service. Car 29705 was found at Wabash, Indiana on March 11, 1973. The black car was photographed in Birmingham, Alabama in March 1969.
(top, Lee Yoder; middle, Jim Thorington)

GATX 81663
DOT105A500W
▼ Also at Birmingham, Alabama was this car, found in May 1970. A 200-ton capacity tank car it was probably constructed in June 1969. The car was one of 15 assigned into the group 81663-81677. *(Jim Thorington)*

GATX 94661
ICC112A340W
◀ Hauling vinyl chloride, the car was seen in August 1968. A 150-ton capacity car, it was placed into the 94650-94706 number group.
(Paul C. Winters)

GCX 731005
ICC111A100W1
▼ An unusual Allied Chemical car seen in Birmingham in April 1971. One of eight such cars numbered 731000-731007, it was assembled by General American in April 1968. The car utilized 440-volt electric heaters (it also appears to have steam heater coils) and was of 30,100 gallon capacity. *(Jim Thorington)*

GCX 741637
ICC112A340W
▲ Probably constructed by Union Tank Car in January 1969, it was spotted in July 1971 while in vinyl chloride service. There were 25 of these 42,000 gallon monsters. They were numbered 741622-741646. *(William Rosenberg)*

MCPX 23020
ICC111A100W1
▲ Not every six axle tank car is enormous in size: here is a 23,000 gallon car built to carry elemental phosphorus for the Monsanto Corporation. It was at Harrisburg, Pennsylvania on June 9, 1973 and was constructed by American Car and Foundry in November 1966. Insulated and having heater coils, it was part of the 37-car group numbered 23000-23036. *(Craig Bossler)*

IAPX 1005
AAR204W
▲ At Portland, Oregon on October 7, 1972 was this car. Although it is a cryogen car, which are covered in the next section of this book, its six axle design places it here. The 24,510 gallon car was assembled in February 1967. It was one of three numbered 1003-1005.
(G. Bolinsky, Jim Kinkaid collection)

IAPX 1008
AAR204W
▶ Another Industrial Air Products car, also at Portland on October 7, 1972. This was a single car order, apparently assembled in February 1968. It had a 24,175 gallon capacity.
(G. Bolinsky, Jim Kinkaid collection)

NATX 37300
ICC111A100W1
▲ The car illustrated here was one of two such cars (the other was NATX 37301). It was located at Thorndale, Pennsylvania on April 25, 1972. Apparently assembled in July 1966 it had a capacity of 33,841 gallons and steam coils. *(Craig Bossler)*

NATX 38017 & 38051
ICC114A340W
▲ ◄ A couple of cars with rather unusual rooftop arrangements. NATX 38017 was found on December 13, 1969. In methyl chloride service it carried 33,857 gallons of this chemical. The other one was photographed in July 1969 and was likely assembled in July 1967. Both cars belonged to the 55-car series numbered 38000-38054.
(middle, Craig Bossler; bottom, William Rosenberg)

RAIX 4309
DOT111A100W1
▼ Although owned directly by Union Carbide when spotted in Irondale, Alabama on May 15, 1993, the car started out as a U.S. Leasing car. It was first listed in the equipment register in July 1970 along with 19 others as RALX 4300-4319 (although the cars were not constructed until November of that year). By October the series became RAIX. *(Ed Trundle, Gib Albach collection)*

SHPX 88362
ICC111A100W1
▼ Another Mobile Oil car. On December 13, 1969 it was spotted in Thorndale, Pennsylvania. Compare the car to the ACFX-marked one on page 114. *(Craig Bossler)*

UTLX 83699
ICC114A340W
▶ True enough, this is a whalebelly car, but its four truck design places it here. The car carries 50,105 gallons of liquefied petroleum gas and was at a railroad show in Cincinnati, Ohio in August 1966. A single car order it was constructed in March 1963 with a tank fabricated by Graver Tank and Manufacturing. *(Paul C. Winters)*

The Exotics: The Cryogen Cars

APRX 2009 & 2011
DOT113C120W
◄ ▼ Two cars carrying liquid ethylene. At top is one spotted at Reading, Pennsylvania on November 3, 1984 and manufactured in January 1976. Below it is another found at Altoona, Pennsylvania in June 1990. It was assembled in March 1976. Both were part of the 11-car series 2001-2011, owned by Air Products and Chemicals Incorporated. They were of 32,900 gallon capacity.
(top, Craig Bossler; middle, Jim Kinkaid)

ARIX 3005
AAR204W
▼ An Airco car at Reading on August 24, 1978. One of 12 cars, it was built in October 1971. The 25,265 gallon car was part of the set 3001-3012 and in liquid nitrogen service. *(Craig Bossler)*

BIGX 1024
AAR204W
▶ Pine Bluff, Arkansas was where this car was spotted on August 22, 1985. In argon service the 16,555 gallon car was first listed in the equipment register in July 1985. It was assigned into the 1019-1024 number series.
(Al Lanier, Jim Kinkaid collection)

GATX 93332
DOT113C120W
▼ In October 1969 General American began to report a new series of AAR class TVI cars numbered as 93325-93326. By January there were ten listed as 93325-93334, to which this car belongs. It was at Belt Line Junction, Pennsylvania on April 25, 1988.
(Craig Bossler)

LTCX 1811
AAR204W
▼ In liquid argon service, this 19,000 gallon car was spotted while at Belt Line Junction on May 16, 1977. Constructed in January 1974 it was part of the 14-car series numbered 1801-1814. *(Craig Bossler)*

NPCX 206
DOT113C120W
▼ Here is a 32,900 gallon car photographed on August 19, 1987 while in Reading, Pennsylvania. Owned by the Northern Petrochemical Company it was constructed in August 1975 and part of the series 201-218. *(Craig Bossler)*

NPCX 220
DOT113C120W
◀ Another Northern Petrochemical Company car at Reading. It was photographed there on October 15, 1988. Assembled in August 1973 it was apparently acquired second hand since it only first began to be listed in the equipment register in April 1978 as one of two cars numbered 220-221. It was also a 32,900 gallon capacity car.
(Craig Bossler)

PTRX 7005
DOT113C120W
▼ There were four of these cars that were first mentioned in the equipment register in October 1970. Numbered 7003-7006 and of 34,150 gallon capacity, it is unknown if they were acquired new or used. This one was found in Little Rock, Arkansas on August 21, 1985. In liquid ethylene service, owner Petrolite's Bareco Division used this substance in the manufacture of commercial waxes and polyurethanes.
(Al Lanier, Jim Kinkaid collection)

PUBX 131
DOT113D60W
▲ A Publicker Industries car found in April 1970. The 15-car set that it belonged to, 128-142, first appeared in the October 1968 issue of the equipment register. In July 1969 ten more were added as cars 143-152. *(Paul C. Winters)*

RAIX 712
AAR204W
▲ Built in April 1976, we're looking at a photograph taken at Jefferson Street in Atlanta, Georgia on March 21, 1992. The 19,400 gallon car was in argon service. It was part of the 701-712 set of cars.
(Ed Trundle, Gib Albach collection)

SUNX 90000
DOT113C120W
▶ Constructed in June 1995 the car was seen at Reading on October 27th of that year. The 33,000 gallon car was one of four numbered as 90000-90003. The Sun Refining and Marketing Company, whose name goes back to 1890, operated these cars.
(Craig Bossler)

UTLX 80032 & 80035
AAR204W
▼ ▼ Some cyrogen cars under lease from Union Tank Car. It would appear that these cars are part of the set UTLX 80010-80039 fabricated in January through March 1963. They were nominal 16,200 gallon cars. Car 80032 sits at Kansas City, Missouri on March 25, 1992. It was assembled in January 1963. At middle is another one found in September 1963. It had been constructed in March of that year.
(top, Jim Kinkaid; middle, Paul C. Winters)

UTLX 80045
AAR204W
◀ More cars were constructed in early 1964 as UTLX 80040-80054. This Airco car was at Pottstown, Pennsylvania on November 8, 1970. It had been assembled in February 1964 and was of 16,306 gallon capacity. *(Craig Bossler)*

UTLX 80046
AAR204W
▲ Another car from the 80040-80054 set. It was found in July 1968, having been supplied in February 1964 too. This car was of a 16,302 gallon capacity. *(Craig Bossler)*

AAR207 Cars, The Hopper Designs

While looking much like cylindrical covered hoppers, these cars were indeed considered tank cars and had tank car specifications. They were primarily used for dry commodites yet had tanks built to tank car specifications. They were rare indeed.

CELX 9302
AAR207A40W6
▲ Assembled in November 1966 by Union Tank Car and put on film in February 1968. It has a stainless steel tank and was part of a Celtran 4-car set numbered 9300-9303. *(Paul C. Winters)*

MOHX 3042
AAR207A40W6
▶ The Monsanto Company owned this car. Besides having a tank car specification, it was listed in the equipment register as an "LO" covered hopper car, indicating the dual nature of this design. Found in Birmingham, Alabama in April 1969 it had been built in October 1966 by Union Tank Car. The car, which was part of the 3040-3045 set, had a stainless steel body and was of a pressure differential design. *(Jim Thorington)*

UTLX 80276
AAR207A40W
▼ A car seen in May 1972. Medusa Cement was one of the oldest cement manufacturers in the United States and the company name can be traced to at least 1929. *(Paul C. Winters)*

UTLX 80378
AAR207A40W
▼ Another found under lease to the FMC Corporation at Lawrence, Kansas, which is where the car was found sitting in November 1988. It had been constructed in February 1969. Sodium carbonate has wide usage, but is primarily useful in glass manufacture. *(Jim Kinkaid)*

UTLX 80457
AAR207A28W
▲ Our last AAR207 car was spotted in August 1972. It was assembled in December 1965 and under lease to a company that was once a major producer of flame retardants and pesticides. *(Paul C. Winters)*

Miscellaneous Cars

Here are three cars that resemble tank car designs presented earlier in this book, but are not "true" tank cars.

DUPX 34500

▶ Compare this car to several on pages 126 & 127: it looks nearly identical, yet it is strictly considered a covered hopper by owner DuPont. It is a 3,000 cu.ft. car assigned to the group 34500-34504, acquired used and which were first listed in the equipment register in January 1976. Baymag is a company that was a major producer of magnesium oxide, an item used in paper, pulp and cements, among other things. *(Paul C. Winters)*

HWCX 87

▼ Even though the car in this photo remembles some kind of special tank car, it is not. It is in fact a car assigned the AAR mechanical designation "LM." These 17,950 gallon cars were used by owner Haliburton Company to move oil rig drilling chemicals. This one was at Birmingham in April 1970. Constructed in 1968 there were 150 of these numbered 50-199. *(Jim Thorington)*

USLX 5013

▶ Looking much like the AAR207 cars just illustrated, it was in fact a covered hopper (note the "LO" stencil). Gunderson constructed 18 of these cars in March 1969 as USLX 5000-5017 under lot 16344, completed at Portland, Oregon. This 5,000 cu.ft. example was at Newark, New Jersey in October 1970. *(Jim Thorington)*